School Broadcasting in Canada

This book describes the origin, growth, and achievements of school broadcasting in Canada. Sections are devoted to the start of school broadcasting in each province, the establishment of national school broadcasts, and the work of the National Advisory Council on School Broadcasting. In the story, the part played by the Canadian Broadcasting Corporation in initiating and promoting the work of teaching by radio and in providing the facilities upon which it is based, is a significant one.

The book is the first authoritative description, by the man largely responsible for its success, of an important and fruitful experiment in federal-provincial co-operation in the thorny field of education. To this co-operation is due the high standard of the school broadcasts which have earned for Canada world-wide recognition and appreciation. The book also describes the international aspects of this cooperation, particularly between Canada and Australia, Great Britain, and the United States.

RICHARD S. LAMBERT (1894-1981) was a biographer, popular historian, and broadcaster. He served for twelve years (1927-1938) as founding editor of the BBC's weekly literary and educational journal *The Listener*. He was also one of the first governors of the British Film Institute. From 1943 to 1960, he was Supervisor of School Broadcasts for the Canadian Broadcasting Corporation. Mr. Lambert is the author of some thirty books on a wide range of subjects including biography, history and travel.

SCHOOL

BROADCASTING

IN CANADA

Richard S. Lambert

CBC SUPERVISOR OF
SCHOOL BROADCASTS 1943–1960

University of Toronto Press

© UNIVERSITY OF TORONTO PRESS 1963
Reprinted in paperback 2016
ISBN 978-1-4875-9278-3 (paper)

Acknowledgments

The author wishes to express his gratitude to the following persons for their help in placing at his disposal records and memoranda regarding school broadcasting, and for checking over appropriate sections of the manuscript:

MR. W. BRUCE ADAMS, Director, Teaching Aids Centre, Toronto Board of Education

MR. DAN CAMERON, Programme Director, CBC Prairie region

MR. KENNETH CAPLE, Director, CBC Pacific region

MISS MARGARET COOKE, Librarian, CBC Toronto

DR. IRA DILWORTH, formerly CBC Director for Ontario, and Director of the CBC Pacific region

MR. G. ROY FENWICK, formerly Supervisor of Music, Ontario Department of Education

MR. C. F. FUREY, Director of Audio-Visual Education, Newfoundland Department of Education

MR. WILLIAM GALGAY, Director, CBC Newfoundland region

MAJOR JAMES W. GRIMMON, Director of Audio-Visual Education, Ontario Department of Education

MR. GERARD LAMARCHE, Supervisor of Adult Education and Public Affairs, CBC French network

MR. A. R. LORD, formerly Chairman, British Columbia School Radio Committee, and Principal of Vancouver Normal School

MR. DOUGLAS B. LUSTY, School Broadcasts Organizer, CBC Maritime region

MISS GERTRUDE MCCANCE, Supervisor of School Broadcasts, Manitoba Department of Education

MR. H. MCNAUGHT, Audio-Visual Branch, Ontario Department of Education

MISS IRENE MCQUILLAN (Mrs. A. Murphy), Supervisor of Music, Halifax City Schools

MR. RICHARD A. MORTON, Supervisor of School Broadcasts, Alberta Department of Education

MISS GERTRUDE MURRAY, Supervisor of School Broadcasts, Saskatchewan Department of Education

MISS MARGARET MUSSELMAN, Director of School Broadcasts, British Columbia Department of Education

MR. GERALD NASON, Secretary-Treasurer, Canadian Teachers' Federation

MR. FRANK W. PEERS, Director of Information Programming, CBC Ottawa

MR. GERALD V. REDMOND, Director of Nova Scotia Tourist Bureau, formerly Supervisor of School Broadcasts, Nova Scotia Department of Education

FATHER AURÈLE SÉGUIN, formerly Director of Radio-Collège, CBC French network

DR. L. P. STEEVES, Director of Audio-Visual Education, New Brunswick Department of Education

MR. F. K. STEWART, Executive Secretary, Canadian Education Association

PROFESSOR MORLEY P. TOOMBS, Department of Education, University of Saskatchewan

The author also wishes to express his sincere thanks to his successor as CBC Supervisor of School Broadcasts, Dr. Fred B. Rainsberry, for his co-operation and encouragement in the preparation of this book.

He also wishes to thank the management of the CBC for facilitating in every way the writing of this book, and for undertaking the responsibility for its publication.

RICHARD S. LAMBERT

July, 1962

Foreword

ALPHONSE OUIMET

With pardonable pride one of our newspaper critics has referred to the audience for school broadcasts in Canada as "the world's biggest radio class-room." Of course, the correctness of this boast is spatial rather than numerical. According to the map, school broadcasts supply a service, every school day, to an area which extends from Newfoundland in the East to Vancouver Island in the West, and from the American border northwards to the main centres of the Canadian Arctic. This service is not dictated from a single national governmental centre, but is democratically decentralized in accordance with regional and provincial needs. Its cost is shared between the broadcasting and the educational authorities. These two factors, applied over the huge area I have indicated, give Canadian school broadcasting a distinctive character which has led to the publication of this detailed study of its growth and operation.

The mainstay of the vast communications area to which I have referred is the network of radio facilities provided by the Canadian Broadcasting Corporation and its affiliated stations. Without them, it can safely be said, there would be no national school radio service, and at best only a partial provincial school radio service for those provinces that could afford it.

The CBC has provided the main impulse, the technical backbone, and the nation-wide scope for school broadcasting in Canada, thanks largely to the enlightened leadership of those who guided the Corporation through the earlier stages of its growth—men like Major Gladstone Murray and Dr. Augustin Frigon (first and succeeding General Managers of the CBC) and Mr. Leonard Brockington and Mr. Davidson Dunton (who presided over the CBC's Board of Governors). These men saw clearly that the national service of broadcasting included the duty of promoting and supporting the educational uses of radio in the class-rooms of Canada.

On the other hand, the actual framing of the educational service itself—which included planning the curriculum, responsibility for programme content, utilization and evaluation of the broadcasts—has lain wholly in the hands of the provincial departments of education, the local school boards, and the teachers. The CBC has limited its function, as it should, to encouragement and promotion, and to ensuring that its facilities were well used in terms of broadcasting technique.

To the educators, school broadcasting has meant the incorporation in their system by voluntary processes of a new tool of learning, "radio." No one would claim that, in the short space of two decades, this incorporation is yet complete. At least, however, the radio receiver has now become an accepted part of the equipment of the average school and, in the subjects for which it is most suited (music, language, literature, and social studies), has made an acknowledged contribution to the effectiveness of teaching. Expenditure on school radio programmes and staff is now a regular part of the budgets of provincial governments, as well as of the CBC. In all provinces local directors of school broadcasting or audio-visual education work hand in hand with specially assigned CBC officials.

Whatever success the work has achieved is due largely to the harmonious co-operation that has been established between educational broadcasters and professional educators. The Canadian constitution clearly distinguishes between federal and provincial jurisdiction in matters of education and communications. It assigns the first to the provinces, the second to the federal government. It is, therefore, no mean achievement to have created, on a working basis, the National Advisory Council on School Broadcasting, which has served with success as a "parliament" of school broadcasting in Canada for nearly twenty years. Tribute should be paid to the skilful leadership of the educators who have sat on this Council, and above all to the merits of the notable chairmen who have in succession presided over its deliberations. The Council has shouldered a responsibility to Canadian educators which the CBC could not have discharged alone.

The publication of this book occurs at a crucial transitional period, which might be called the "watershed" between sound broadcasting and television in the educational field. It will be necessary henceforth for the educators to accommodate two media, instead of one, in their system of instruction. The change is bound to produce dislocations and fresh problems. For there are profound differences between radio and television, particularly in matters of technique, facilities, and costs. Experience has already shown that new approaches, new methods, and, above all, new financial arrangements will be needed if school television is to mature in Canada to the same level as its precursor, school radio. The last chapter of this book, describing the early experiments conducted with school television in this country, also forecasts some of the changes likely to be called for. The responsibility for developing school television must be spread over a much wider field than heretofore. Also, much more money will have to be found through a greater contribution of funds from public educational sources.

This record of the "sound" era of school broadcasting can do much to awaken educational and public opinion to the nature of the problem which we all face in the coming of school television. Many lessons can be learned from the history of sound school broadcasting which are applicable to the new situation. One thing is certain: there will be the same need of co-operation and mutual understanding between educators and broadcasters. It is also highly likely that, for many years to come, sound broadcasting and television will coexist and supplement each other as essential tools for up-to-date teaching.

The compiler of this history, Rex Lambert, has for 17 years served in the double capacity of CBC Supervisor of School Broadcasts and Secretary of the National Advisory Council on School Broadcasting. This experience has given him an insight into the problems and attitudes of both parties, and has enabled him to gain the confidence of both. In compiling this record, he has acknowledged his debt to the departments of education and other Council members for their generosity in making accessible to him their papers relating to the early history of school broadcasting, and for checking the drafts he prepared relating to the beginnings of the work in each province. The CBC is happy to have enabled Mr. Lambert, by prolonging his service with the Corporation as consultant, to complete the record. We believe it will hold much of interest not only to Canadians, but to educators and broadcasters in all parts of the world as well.

Mr. Lambert came to his job to face a situation in which provincial educators were understandably wary of co-operation with a federal agency in their exclusive field of jurisdiction. Displaying a remarkable ability to work with people who were autonomous in their own fields, he made a going thing out of the National School Broadcasts. Largely because of his interests in such things as music, drama, art, history, and story-telling, he brought an entertaining variety to programming in the educational field. An astounding literary output of his own, ranging from books for children and books on Canadian history to books on wildlife, stands as testimony to the fact that he contributed to his work in much more than an administrative way.

Throughout his service with the Corporation he held the deep conviction that the work of his department was at least as important as anything else being done by CBC. It is difficult to disagree with him.

Contents

ILLUSTRATIONS

between pages 68 and 69

The National Advisory Council on School Broadcasting in session in 1959

Children assisting in a British Columbia junior music school broadcast, "Alice in Melodyland," in 1941

A demonstration of the School of the Air of the Americas held at Harbord Collegiate Institute, Toronto, in May 1941

Planning the first music appreciation series in the Ontario school broadcasts, 1943

Miss Irene McQuillan and a group of children taking part in a junior music broadcast to Maritime schools

Planning a national school broadcast in 1951

Mrs. Dorothy Adair, Grade I teacher at Givins School, Toronto, listens with her class to "Kindergarten of the Air" in 1950

A class practising art expression in the CBC's Winnipeg studios during a broadcast of "It's Fun to Draw"

School Broadcasting in Canada

I. Introduction

Still within memory are those exciting days after World War I when sound broadcasting burst upon a receptive and appreciative Canada. Majority opinion was quick to grasp its possibilities for popular entertainment, as well as for commercial and political propaganda. On the other hand, the minority had equally high hopes of what the new medium could do for education—at the university level, in the home, and especially in the school class-room. Enthusiasts spoke of a revolution in curricula and teaching methods; of a levelling of educational opportunity between town and country; and of the opening of "windows of imagination" in the walls of the average class-room to receive vivid, direct impressions and experiences from the great world outside.

None of these educational visions was fanciful. All have since been, to a considerable extent, realized. The only miscalculation—if we may so term it—was about the speed with which changes could be introduced into that most conservative element in our social framework, the school system. It is not enough to demonstrate the desirability of any new educational tool, and expect that alone to ensure its early adoption. That desirability has also to overcome the handicaps of local inertia, competition from other equally desirable tools (such as films and records), and the financial and technical problems involved. In Canada, with its vast size and educational diversity, acceptance of school broadcasting was bound to come gradually and piecemeal.

None the less, the progress made in little more than twenty years has been remarkable when compared with parallel progress in the United Kingdom (with its denser population), the United States (with its much greater financial resources), and the other countries of the Commonwealth. There are four distinctive features of this progress.

First, in less than twenty years the radio receiver has become an accepted part of the equipment of the average class-room, at the primary, junior, and senior elementary grade levels, through English-speaking Canada. This is less true at the high school level, although there the tape recorder has partly supplied the deficiency. It is not yet true of the schools of French-speaking Canada, except in the province of New Brunswick.

Second, a school broadcasting service has been established over a wider geographical area than anywhere else in the world. It extends from Newfoundland on the Atlantic to Vancouver Island on the Pacific, and from the American border northward to the Canadian Arctic. Throughout this area the schools have available to them, through the CBC network, thirty minutes of school broadcasting every school day from October to May. At the same time, this service of programmes is not centralized or standardized, but is varied in accordance with local curricular needs.

Third, the standard of programming has been consistently maintained at a high level, as is attested both in the reports of teachers to their local education authority and by the many awards bestowed on Canadian school broadcasts over the past 20 years by the internationally known Institute for Education by Radio and Television at Columbus, Ohio. In general, Canadian school broadcasts rely upon dramatized or semi-dramatized methods of presentation, rather than on straight talks or lectures. Their aim has been to supplement rather than to substitute for the class-room teacher, and to enrich his teaching capacity and the learning experiences of his students.

Fourth, while the majority of the programmes have been tied closely to the curricular requirements and teaching techniques of the individual provinces of Canada, approximately one-fifth of all the school broadcasts have been devoted to the wider aim of strengthening Canadian unity and making the younger generation more aware of the responsibilities of Canadian citizenship. At the same time, Canadian school broadcasts have also drawn heavily on the resources of other countries, especially Britain, the U.S.A., and the Commonwealth. Canada originated and has for years maintained programme exchanges with other Commonwealth countries, especially Australia, by which an international flavour has been imparted to Canadian school broadcasting.

CO-OPERATION BETWEEN EDUCATORS AND BROADCASTERS

These distinctive features of Canadian school broadcasting are the result of a policy of close co-operation over the years between the educational and broadcasting experts of the country. The existence of the Canadian Broadcasting Corporation as a federal agency devoted to developing public service broadcasting for the nation as a whole has been essential to the successful growth of school broadcasting. As early as 1939, only three years after its start, the CBC carried out a nation-wide survey of school broadcasting in Canada (see chapter III) which indicated where its support could best be given. Without the CBC there might indeed have been a sporadic and unequal growth of school broadcasts in this or that area of Canada. But it would hardly have been possible to develop a balanced provision for the whole country which, at the same time, promoted national unity and protected local diversity.

Equally important has been the close involvement in the work of the recognized education authorities of Canada, the departments of education and the school boards, which has kept the system of school broadcasting on a practical basis, and has encouraged school teachers to use the programmes. (This is important because, in Canada, listening to school broadcasts is a voluntary activity, and teachers therefore use the broadcasts only if they are justified in terms of class-room value.)

FINANCE OF SCHOOL BROADCASTING

Only in Canada do the school authorities share with the broadcasting authority the cost of school broadcasting. Most school broadcasts are provincial or regional in scope. Their cost is shared with the CBC on the basis that the departments of education pay all direct programme costs (scripts, copyright and research fees, actors, musicians, etc.) while the CBC provides free of charge the production and administrative staff, transmitter time, studio facilities, and network charges. As some indication of how this cost-sharing works out in practice, during the year 1957–58 the CBC's expenditure on school broadcasts (on a cost-accounting basis) was $214,820.

Table I gives figures of the budgets allocated to school broadcasting

TABLE I

SCHOOL BROADCASTS BUDGETS

Province	1948–49	1961–62
British Columbia	$16,910	$20,150
Alberta	20,000	30,000
Saskatchewan	15,000	28,240
Manitoba	10,000	30,500
Ontario	19,000	33,000
Quebec (Protestant)	140	500
New Brunswick	4,000	7,000
Prince Edward Island	600	1,500
Nova Scotia	4,000	7,000
Newfoundland	nil	19,500

by the ten departments of education at two periods, 1948–49, and 1961–62, for the sake of comparison. The figures exclude salaries of departmental personnel and grants made to school boards to aid them in obtaining receiving equipment.

A third partner in school broadcasting is the group of private stations affiliated to the CBC which carry the school broadcasts voluntarily, as a part of their programme schedule. School broadcasts have never (down to 1961) been placed in "reserve time" (i.e., been made part of the contractual obligation of affiliated stations). However, a large number of private stations have regularly carried the programmes as a public service. The value of air time provided by them may also be reckoned as a considerable item in the cost of school broadcasting.

MACHINERY OF CO-OPERATION

From its inception in 1936, the CBC indicated its readiness to place its facilities, as far as feasible, at the service of the departments of education. Furthermore, it undertook *not* to present over its facilities any school broadcasts that had not received the approval of the appropriate departments of education. On their side, of course, the departments were not limited to use of CBC facilities only. A department of education can (and sometimes does) satisfy a part of its school broadcasting requirements by making special arrangements with private radio stations, without the intermediacy of the CBC.

The formula upon which co-operation between the CBC and the departments of education is based is as follows. The education authority takes responsibility for the planning and content of the broadcasts. The CBC is responsible for their presentation and distribution. Vague as this formula may seem, it has stood the test of practice well. It has been found to afford a workable basis for establishing good provincial–federal relations in this somewhat "touchy" educational sphere. Throughout the past twenty years the CBC has co-operated harmoniously with the departments of education on this basis, and the broadcasts they have jointly provided have proved, both in form and in content, satisfactory to the majority of Canadian teachers, students, and parents. This in itself is no mean achievement.

Co-operation between educators and broadcasters takes place at three levels—local (municipal), provincial or regional, and national. At the local level, the co-operating parties may be a local school board and a local radio station, either privately or publicly owned. At the provincial level, the co-operation is between one or more departments of education and the appropriate CBC regional authority, which sets up the network of stations required to give the desired coverage. Four-fifths of all the school broadcasts in Canada are conducted at this level. At the national level a different situation prevails. Here special machinery for co-operation is necessary because, under the British North America Act, education is exclusively a function of provincial government, and no federal education authority exists. Each provincial department of education is autonomous in all matters relating to schools in its area. This prerogative is jealously guarded in all ten provinces.

Broadcasting, on the other hand, is constitutionally a function of the federal government. The CBC, as a federal agency, controls the network production facilities necessary for a national school broadcast, and cannot divest itself of final responsibility for whatever goes on the air over its facilities.

At the national level, co-operation between the CBC and the departments of education has been achieved through the National Advisory Council on School Broadcasting, a committee set up in 1943, by agreement between the Canadian Education Association and the CBC, to represent not only the departments but also other national bodies concerned with education. The functions and working of the National Advisory Council are fully discussed in chapter v. Here again the formula for co-operation is the same as that for local and provincial school broadcasts. The Council is responsible for the planning, the CBC (through its School Broadcasts Department) for the presentation of the programmes. In the case of the national school broadcasts, by agreement between the educators and the CBC, the full cost of the programmes is borne by the latter.

THE ROLE AND TRAINING OF THE CLASS-ROOM TEACHER

The part played by the class-room teacher is vital to the success of school broadcasting. He is centrally involved in the making of the programmes, and without his goodwill and interest the broadcasts may not reach the students at whom they are aimed, or be suitably utilized in class work.

In the planning of the broadcasts, most departments of education make use of committees of teachers, or of subject-specialists, to advise them. When the plans are ready, and are turned over to the CBC for presentation, selected class-room teachers are employed as consultants, script-writers, and often as performers. Where the programmes are of the lesson type (as is common in provincial school broadcasts) outstanding teachers naturally deliver them with greater ease and skill than could professional broadcasters. Of course the elements of "personality" enters in, and it is not every good teacher who can adapt his class-room techniques successfully to broadcasting needs.

When the programmes are presented in dramatized form, the case is different. Here the professional teacher often cannot do as well as the professional actor. In addition, the CBC finds itself bound by the regulations in its contracts with the various artists' unions, which do not permit "amateurs" to work along with professionals, except in special cases. Teachers, however, can be employed on these broadcasts provided they are performing a "teaching" rather than a "performing" role. Writing dramatized scripts, too, requires professional skill, which it would take the average teacher too long to acquire. A number of ex-teachers, however, have turned to script-writing for a livelihood, and these draw upon their past class-room experience to enable them to turn out dramatized scripts of high quality.

Even more important than the teacher's role in the broadcasting studio is his role at the receiving end, in the class-room. Here it is not excessive to say that his attitude may make or mar the value of school broadcasting.

It is generally admitted that school broadcasts cannot take the place of the class-room teacher. Teaching is a complex and highly professional art dependent on the skilful manipulation of an interplay between two persons—teacher and student. So far, this interplay is beyond the capacity of any electronic device. School broadcasts are therefore no more than "aids to teaching," that is, tools which extend the scope and effectiveness of the teacher's work. In this respect radio and television receivers rank with phonographs and records, film projectors and motion pictures, tape recorders, and other similar "teaching aids."

Properly used by skilled teachers, school broadcasts contribute substantially to education in several ways. They give imaginative enrichment to the students' study of certain subjects, particularly history, geography, music, science, and literature. They help strengthen their motives to study, sharpen their capacity to take in information, and stimulate them to further pursuit of knowledge. School broadcasts extend the range of students' experience far beyond their class-room walls. They give greater reality to geography lessons through broadcast visits to foreign lands and peoples. They make possible natural science field trips to places outside the local environment (as in the popular national series "Voices of the Wild"). They also help to make world events of importance real and understandable in the class-room. Another outstanding gift of radio (and television) is its power to bring to the microphone the interesting personalities of the day, experts and leaders in all forms of endeavour.

Utilization and Evaluation

But, as we have said, however carefully school broadcasts are planned and put on the air, they cannot fully be effective unless suitably received and utilized in the class-room. Mere passive listening is usually disappointing. All good school broadcasts are deliberately planned so as to leave something positive for the teacher to do to complete their effectiveness. Effective utilization of a school broadcast depends on the teacher's ability to integrate the programme into the lesson he is giving. This involves preparation before the broadcast, and "follow-up" activity after it with the students. The truth of this is illustrated in the following incident. Two teachers in different schools, having received the same broadcast in their class-rooms, sent in reports on the result to their department of education. One wrote: "Not worthwhile. After a few minutes the students' attention wandered, and they became bored. At the end they voted that we discontinue listening to school broadcasts, and use our time for textbook study." The other said: "My students followed the broadcast with rapt attention. They learned from it, in those twenty minutes, more than I had been able to put into their heads during a whole term's work." The difference in result reflects the difference in the two teachers' skills in utilizing the programme.

A prerequisite to successful utilization is a supply of adequate information to the teacher about forthcoming programmes. From the early days of school broadcasting, both the CBC and the departments of education fully realized this. From 1940 onwards they issued regular programme guides for distribution to all interested teachers. These guides included a calendar of dates of the years' school broadcasts, titles and summaries of the individual programmes, and, often, pictures, diagrams, and maps needed for effective utilization.

Teachers were also asked, from the outset, to evaluate the broadcasts that they used in their class-rooms. Most departments of education made provision for regular reports to be sent in to them during the school year, and from these they compiled a detailed record of progress. However, in the case of the national school broadcasts, covering all ten provinces, the procedure for evaluation proved much more complicated, and for a variety of reasons (as will be seen in a later chapter) a less effective result has been achieved.

Training teachers in the use of school broadcasts has been a gradual process, always improving, but never reaching completion. Many older teachers looked down on radio as an "educational frill." The obvious place to begin such training, therefore, was with the new teacher about to enter the profession through the normal school or the college of education. In these institutions, however, the curriculum is heavily overloaded, and emphasis is laid on traditional teaching techniques rather than on the new "teaching aids." In some teacher-training institutions lip service only is paid to school broadcasting, in the form of an occasional lecture or demonstration by the departmental officer concerned, or a visit to a local broadcasting studio. The young teacher is indeed exhorted to remember that school broadcasts are provided to help his students in certain subjects; but he is left to find out for himself, by experience, the best ways of making use of them.

The deficiency is partly offset by the provision, in most provinces, of summer courses, or week-end institutes, which teachers already in service are encouraged to attend. At some of these courses most valuable instruction is given in the use of radio and film, and practical opportunities furnished of handling equipment, planning programmes, practising speaking and writing scripts, and discussing methods of utilization and evaluation of school broadcasts. In certain large urban centres directors of audio-visual aids have been appointed, whose function is to stimulate teachers to use the newer aids. Toronto, for example, has for many years had a Teaching Aids Centre, with a director and staff, which trains personnel, circulates information, and demonstrates new types of equipment. Also, in the Ryerson Institute of Technology, the Ontario Department of Education provides a variety of electronics training courses which fits students to enter the radio and television professions.

What sort of school listens to school broadcasts? Not necessarily the big city school, which already enjoys the cultural advantages of an urban environment. Not necessarily a school with many class-rooms wired to a central sound equipment. More likely, a school in a smaller town, where life moves at an even pace, without many outside influences. Or even a "little red schoolhouse" of the traditional type (of which many are left), buried in the heart of the country, attended by farm children, and possessing few contacts with the world outside. Here is where school broadcasts have their biggest opportunity, and are most warmly appreciated by teachers and students. To these we may add the children of Canadian service men serving in Europe or at isolated stations in Canada (who receive the programmes on delayed transcriptions), Eskimo children in the Arctic, Indian children attending reservation or mission schools, children of well-to-do parents enrolled in private schools, and children of trappers and loggers who depend on correspondence lessons and radio for their education.

To receive school broadcasts in its class-rooms, a school must acquire one or more receivers. They are usually provided out of public funds, or sometimes by private assistance. Table II shows the extent to which departments of education assisted their schools to acquire receivers during the formative period when school broadcasting was growing.

TABLE II

DEPARTMENTAL ASSISTANCE IN 1948–49 FOR PURCHASE OF RECEIVERS

Province	Grants
British Columbia	up to 50 per cent of cost
Alberta	25 per cent of cost
Saskatchewan	40 per cent of cost
Ontario	30–70 per cent of cost
Quebec (Protestant)	75 per cent of cost
New Brunswick	50 per cent of cost
Prince Edward Island	50 per cent of cost

Two provinces, Nova Scotia and Manitoba, followed the policy of making little or no grant for this purpose. On the other hand, Newfoundland, coming later on the scene, found it necessary, in many cases, for the department to provide schools directly with receivers.

Today, many departments of education merely list radio receivers as school equipment which, along with films, projectors, slides, and other aids, ranks for grant aid.

Many schools in this country are still inadequately equipped, and even the best are limited by their physical resources. Radio has a great opportunity to reduce the inequality between schools, especially between urban and rural schools, for a school broadcast can be shared by rural and urban children alike, regardless of the limitations of their respective buildings and equipment. Unfortunately, it is still common to find class-rooms, especially in rural areas, where the teacher and students have to rely on small mantel-type radios, even to receive music appreciation broadcasts, which are the most popular of all the programmes provided. There is a tendency not to replace old and obsolete receivers provided in earlier days. This accounts, in part, for many of the complaints of unsatisfactory coverage received from schools in remote areas.

Size of Class-room Audience

Measuring school broadcast audiences is a very different matter from estimating the number of listeners to home programmes. Telephone and panel surveys are not applicable. The CBC is therefore largely dependent upon the information supplied to it by departments of education. Each department keeps an accurate record of the pupils listening to its own school broadcasts, but the method of presenting that information to the public varies from one province to another. It has proved difficult to collate that information upon a uniform national basis.

Until 1951 the best method of estimating the size of the school listening audience was through the statistics compiled annually by the Department of Transport, Ottawa, showing the number of applications made each year by schools for free radio receiving licences. For example, the figures for 1948 showed that out of 22,257 English-speaking schools in Canada, 5,861 had applied for free licences. The departments of education estimated that those licences covered 20,146 class-rooms with 429,087 pupils. For the year ending March 31, 1951 (when radio receiving licences were abolished), 8,252 free licences were issued to schools. This represented a remarkable increase for the three-year period. In 1951 the total number of English-speaking schools (according to the CEA) was 23,139; it appeared therefore that the percentage of radio-equipped schools was then 35.6. The highest percentage was 57.1 in British Columbia, then 45.8 in Ontario,

37 in Quebec (Protestant), 34 in Saskatchewan, 31.9 in Nova Scotia, 31.7 in Manitoba, 29.4 in Alberta, and 28.3 in New Brunswick. The estimated size of the total school audience on any one occasion was 600,000 students.

Shortly after this date receiving licences were abolished, and thereafter no further statistics of schools receiving free licences were kept. In 1956, at the request of the National Advisory Council on School Broadcasting, this writer compiled from figures supplied by departments of education in their annual reports a conservative estimate of 32,518 school class-rooms equipped with radio receivers (in four provinces this included the number of outlets for public address systems). Of this number, 15,743 (48.4 per cent) were in Ontario, 4,079 (12.5 per cent) in Saskatchewan, 3,500 (10.7 per cent) in Manitoba, 2,900 (8.9 per cent) in Alberta, 2,315 (7.1 per cent) in British Columbia, and 1,600 (4.6 per cent) in Nova Scotia. These figures suggest a rise in school listening in the Prairies, and a relative fall in British Columbia and Nova Scotia.

A further clue to the size of the audience can be found in the experience of Canadian Industries Limited which, in September 1959, published a four-page pictorial folder designed to supplement the national school broadcast series "Voices of the Wild" (grades 4–6) presented during October and November over a CBC network of 65 stations. Teachers were invited to apply, through an order form in *Young Canada Listens,* direct to C.I.L. for copies of this folder. As a result, over a quarter of a million copies were distributed to 8,653 class-rooms throughout Canada, indicating that approximately 34 per cent of all students in grades 4, 5, and 6 listened to these particular broadcasts—which, of course, formed only a small fraction of the programmes provided for the various grades from 1 to 13.

THE HOME LISTENER AND THE ROLE OF THE PARENT

Although intended primarily for use in school class-rooms, school broadcasts have always enjoyed a large following among adults who listen at home or in their automobiles. From the earliest days of school broadcasts, parents expressed favourable views about the school broadcasts, and voiced a demand that schools be given the facilities to hear them. The lead was taken by local Home and School associations which in many cases raised the matter in discussion with school principals and class-room teachers. Some associations went so far as to raise funds for equipping their local school with receivers.

The Canadian Home and School and Parent–Teacher Federation nationally, and its constituent provincial federations provincially, gave the CBC unremitting support in its efforts to acquaint the departments of education and the school boards with the potentialities of school broadcasting. At the annual conferences of these bodies, frequent resolutions were passed giving prominence to the need for school broadcasts. All these efforts have exercised a powerful influence on Canadian educators in favour of this new medium of education, and in particular have caused many school trustees and school board officials to view them with approval. Incidentally, the Home and School Federation has from the outset been actively represented on the National Advisory Council on School Broadcasting.

As the result of a request made by the Federation, the CBC decided to publish a special edition of its programme guide to school broadcasts, *Young Canada Listens*, for the benefit of the home audience. A considerable proportion of the mail reaching the CBC School Broadcasts Department comes from parents, particularly mothers. The letters invariably stress the value of school broadcasts to Canada as a whole. Most parents say they wish there had been such a service when they attended school. It is interesting to note that school broadcasts are the only portion of class-room activities which parents can share directly with their youngsters.

PROBLEMS OF SCHOOL BROADCASTING

For many years the main problem with school broadcasts was to secure adequate station coverage. Today, however, all but a very small portion of the country is well served, and the few remaining black areas are likely to be covered during the next year or two, when the CBC network is reorganized.

The time of day at which school broadcasts go on the air has always been a compromise between the needs of the schools and the flexibility of the CBC network. Wide local variations occur in school hours, including the opening and closing of the school day and the lunch and recess periods. The CBC has always done its best to meet the educator's wishes, and for many years an early afternoon period has been favoured in the West, while the East until recently preferred morning periods.

The greatest limitation on the use of school broadcasts has been the fact that the programmes are heard only once, so that, if a specific period cannot be fitted into the school time-table, then the broadcast scheduled for that period is lost. Hence there has been voiced by the educators a demand either for a repeat period at another time or for the provision of taped or recorded versions of the programmes.

Cost is the main factor which prevents the distribution of school broadcasts in recorded form. Union and copyright restrictions require the payment of substantial extra fees for recorded programmes; and so far no education authority has been found willing to shoulder these costs. On the other hand, the CBC, existing for broadcasting purposes, cannot undertake to provide schools with records and tapes at its own expense. Co-operative action between several departments of education might solve the difficulty, but only if the departments were prepared to guarantee costs or make bulk purchases of such tapes or records.

Cost is also a factor which limits the progress of school broadcasting in general. The budgets, both of the departments of education and of the CBC, do not allow for much expansion of school broadcasting in the near future. Yet there is considerable need for more staff to undertake increased "field work" among teachers to acquaint them better with methods of utilizing school broadcasts.

At the present time, programme costs are rising, owing largely to higher performers' fees. One way to offset these rising costs would be to develop a greater interchange of programmes between individual provinces. Much has already been done in this field through the co-operative programmes of the three Maritime provinces, and the four Western provinces. However, there has been comparatively little programme interchange between East and West. In such subjects as music appreciation and French language instruction, where different broadcasts are given in different parts of the country, it might be possible to achieve a greater pooling of resources, with a consequent saving in costs.

In spite of its solid achievements over the past two decades, school broadcasting has still some distance to go in establishing its own importance in the minds of those who make policy in the departments of education. At the ministerial level constant changes take place, and it is common enough to find a new minister of education who, on taking office, has heard little or nothing of school broadcasts. There seem to be, then, wide fields of business and public life into which notice of the new educational medium has not yet penetrated. Publicity for school broadcasting still needs to be substantially widened.

COMING OF CLASS-ROOM TELEVISION

Since 1954 the progress of school (radio) broadcasting has been affected by the rise of television and by the need for exploring its class-room possibilities. The well-knit Canadian provision for radio cannot by itself assimilate the new medium, which involves much greater costs, and calls for increased activity at the local (municipal) as well as at the provincial level.

With the support of the CBC, the National Advisory Council on School Broadcasting has conducted several experimental series of national school telecasts. Certain departments of education and individual school boards have also conducted their own experiments, using CBC facilities in some cases, and private-station or closed-circuit facilities in others. The general conclusion of all these experiments has been favourable to television as a teaching aid in the class-room, but the results have not been sufficiently specific to convince the education authorities generally that they should embark upon large expenditures in this area. The CBC, as before in the case of radio, has taken the lead in demonstrating what possibilities television holds for education, but is also prevented by budget policy limitations from embarking on large scale programming in school television.

Meanwhile, the example of Britain (eight half-hour programmes per day for schools) and the United States (over 50 educational television stations, with generous financial support from the Ford Foundation through its Fund for the Advancement of Education and from other bodies) has shown what could be done in Canada, were resources and enthusiasm sufficiently developed. The encouragement given by the Board of Broadcast Governors, the awakening interest of the Canadian Education Association, the establishment of the Metropolitan Educational Television Association in Toronto, and (it is hoped) of counterparts in other cities, all suggest that progress in the next five years will be substantial. Some intermeshing of Canadian school television efforts with American television school plans would seem likely.

PART ONE

The Formative Period in Provincial School Broadcasting

II. The Maritimes

NOVA SCOTIA POINTS THE WAY

To Nova Scotia goes the credit of starting the oldest continuous system of local school broadcasting in Canada. The impetus was given by Dr. Henry Munro, who in 1926 left his position of Professor of Political Science at Dalhousie University to become Superintendent of Education for Nova Scotia. For many years before, the school curriculum of the province had been slanted heavily in the direction of science and mathematics, and Dr. Munro was anxious to redress the balance in favour of the humanities, especially English and history. During the earlier years of his administration he introduced many innovations, including an annual summer school for teachers, a teachers' pension scheme, and the use of films, records, and broadcasts to supplement class-room work.

Dr. Munro was keenly interested in the commencement of school broadcasting in Britain in 1927. In that year an experiment, financed by the Carnegie United Kingdom Trustees, was carried out in the schools of Kent which led the British Broadcasting Corporation to begin its regular series of programmes to British schools. On reading the report of the Kent experiment, Dr. Munro at once conceived the idea of repeating it in Nova Scotia. He was assisted by a young graduate of Dalhousie University, Victor Seary, who became Secretary of the Nova Scotia Department of Education in 1926, and was in charge of the Department's official *Journal of Education*.

Dr. Munro agreed with the conclusions drawn from the Kent experiment, that the function of radio in the school was "to provide imaginary experiences for children on which their own teachers may profitably build," rather than to provide any kind of substitute for teaching. He approached Major W. C. Borrett, Managing Director of radio station CHNS in Halifax, to see if the station would co-operate with the Department by providing free afternoon time for an experimental broadcast to the schools of Halifax and surrounding areas. The purpose was stated as "purely and simply a test to determine the practicability of applying the English scheme to Nova Scotian conditions." "The Department is firmly convinced," added Dr. Munro, "that radio may become a useful agent, educationally, in this Province." Major Borrett responded enthusiastically, offering the Department the free air time requested and indicating his expectation that the signal strength of CHNS would shortly be increased to give it coverage over most of the province.

First Experimental Programme

Accordingly, on Monday, March 19, 1928, the first experimental school programme was presented over CHNS from 2:00 to 4:00 P.M. Victor Seary, who organized the programme, was able to draw on much excellent local talent, including Professor C. H. Mercer of Dalhousie University, a life-long enthusiast for teaching languages by audio-visual methods and especially radio; Chesley Allen, Superintendent for the School of the Blind, and well known for his lectures and readings on wild life; Dr. Archibald MacMechan, an outstanding teacher and Nova Scotia historian; and the dramatic and musical talent of the King's College Players. The programme presented on March 19 consisted of the following nine items: introductory remarks by the Superintendent of Education; a French lesson conducted by Professor Mercer of Dalhousie University; a scene from Sheridan's "The Rivals" by the King's College Players; a scene from "The Merchant of Venice" by the pupils of St. Patrick's (Girls') High School; a nature talk by Mr. E. Chesley Allen, Superintendent of the School for the Blind; musical selections by the Harmonica Band of St. Patrick's (Boys') School, under the direction of Mr. Cyril O'Brien; a lesson on the correct use of English by Miss M. A. Beresford of Bloomfield High School; an address by Dr. Archibald MacMechan; and closing remarks by the Superintendent of Education.

Report forms were sent out to teachers covering an estimated 10,000 school pupils, and the replies were highly enthusiastic. The Nova

Scotia *Journal of Education* declared that "almost every report on the test expressed the approval of the educational use of radio, and the majority of the reports asked that continued use be made of it in the schools. Almost all adverse criticisms had to do with the weakness of the signals." Incidentally, the programme was picked up in a number of places outside the province, including Saint John (New Brunswick), and Charlottetown (Prince Edward Island).

Regular Programme

Encouraged by this reception, Dr. Munro proceeded in the fall of 1928 to establish school broadcasting on a regular basis of a two-hour programme every Friday afternoon from 2:00 to 4:00 P.M., from October to May. The two hours were divided into 15-minute periods covering six different topics interspersed with brief musical interludes and announcements. All the subjects were treated as supplementary to class-room teaching, using the techniques in the British school broadcasts. They included talks, readings, dramatizations, French language lessons, and vocal and instrumental music, including performances by high school orchestras. Particular attention was paid to dramatizations of Canadian history and geography, world history and geography, and outstanding works from English literature. The programmes also included instructions in French, talks and dialogues on vocations, art and art appreciation, music, science, travel, agriculture, nature study, and folk tales. Special programmes were presented on particular days of national importance, such as Armistice Day and Empire Day.

The programmes were broadcast partly by selected teachers and partly by experts outside the class-room. By 1928 CHNS had increased its signal strength to 1,000 watts, and gave coverage over the mainland of Nova Scotia in the area extending from Lunenburg in the south to Amherst in the north. Use was also made of CHNS remote control facilities to include a series of actuality broadcasts, which enabled school listeners to visit industry plants, railroad and shipping terminals, the airport, the fishing schooner, and similar places of local interest.

In 1930 the schools of Cape Breton Island had their first taste of school broadcasting when the educational series was launched from CJCB, once every two weeks. In 1936 and 1937 these school broadcasts were given weekly.

In these days comparatively few schools possessed their own radio receivers. As early as 1930 the Margaret King School at Pugwash Junction was equipped with a powerful radio receiver of its own. But most schools rented their receivers from local dealers; at other times teachers or pupils brought their own receivers into the class-room. The Department also gave permission to teachers and students to listen in private homes. Gradually a number of schools gained the consent of their local Boards to purchase receivers. In the first four years about 100 schools were equipped in this way. Unfortunately, as a rule, the receivers purchased were of the small mantel type which gave a limited quality of reception for class-room purposes, especially for music programmes.

Through the Nova Scotia *Journal of Education,* the Department of Education issued detailed instructions to its teachers on the class-room utilization of the school broadcasts. Dr. Munro's directive ran as follows:

Classes should discuss the subjects of the radio talks both before and after hearing them. References should be made to maps, pictures and diagrams. Books appearing on the subjects should be available for the private reading of pupils who wish to learn more about what they have heard over the radio. The atmosphere of the classroom during discussion of radio talks should be kept as informal as possible and a premium placed on individual and voluntary research by pupils. . . . The topics discussed in the broadcastings should be applied to the ordinary school subjects so that the programs will serve to supplement the regular texts. Some form of written work should also follow the reception of each program, and if possible, this should be voluntary and not made into a rigid classroom activity.

Complaints were received from some teachers that the broadcasts were too advanced for junior pupils. However, as Dr. Munro wrote, "it is not the intention of the Department that an entire school of eight or ten grades should listen to a whole program . . . best results will be obtained by pupils of the junior and senior high school grades."

Dr. Munro was particularly interested in keeping alive the local and national songs of the people of Nova Scotia. He saw a way to do this through cultivating by radio an appreciation of folk songs in the schools of the province. To develop this interest in music, Dr. Munro turned to Mr. B. C. Silver, subsequently Superintendent of Schools, Halifax County, Nova Scotia, the Supervising Principal of Wolfville High School, who was also pipe-organist of the Baptist Church of that city. Mr. Silver, a native of Lunenburg, had developed an excellent school orchestra drawn from students from grade 5 to grade 11. He directed a band of some thirty instruments, including strings, woodwinds, and percussion. The Wolfville school orchestra contributed to the Nova Scotia school music broadcasts and thereby encouraged music making in other schools of the province. They rehearsed in school for many hours before each broadcast, but went on the air in the Halifax studio after a warm-up playing a single number. Their first broadcast drew a letter of congratulation from a group of lumbermen in northern New Brunswick who had been thrilled to pick up the performance by chance in their camp.

An amusing *contretemps* illustrates the difficult conditions under which school broadcasting took place in these early days. Reports received from many Halifax city schools immediately after one of the programmes made it evident that for a few minutes many listening students and their teachers believed that an explosion had wrecked the studios of CHNS. The incident had resulted from the solicitous action of a station employee. While a speaker was presenting a talk in the main studio, a school choir and school orchestra were assembling outside the studio door, waiting to be admitted, when the talk ended, to give their own programme. The employee had to enter the studio on some technical business, and attempted to quiet the waiting performers as he opened the studio door. Putting his fingers to his lips to request silence, he backed through the door into the studio and immediately tripped over the first row of chairs, taking with him in his fall two additional rows. The crash which followed was loud enough to drown out the speaker at the microphone completely and to leave the listening audience stunned as if by a thunder clap.

A similar occurrence took place when, during a music recital, a soprano singer "blasted" the microphone with a high C of such intensity as to blow a power tube. The engineer in charge promptly replaced the tube, but the soprano then blasted it a second time. The station went off the air long enough for the engineer to hurry downtown in a taxi to get another replacement!

An amusing incident of another kind is recalled by Victor Seary in connection with a talk given in February 1932, on the occasion of the fiftieth anniversary of the death of the famous surgeon Lord Lister.

The speaker was Dr. John Stewart, Dean Emeritus at Dalhousie University and, as I recall, the last surviving member of Lister's classes at Edinburgh. He was a great man in his own right, and a noble figure, very tall and erect, with a snowy beard. I stood in very considerable awe of him. He had never broadcast before, and I explained how far he should stand from the mike—that he should "talk" rather than "speak" as though he were conversing with a young person in his own home, how he should drop his pages to avoid rustles and explosions on the mike, etc. When he finished an excellent talk, I walked over to him to take him off the air, and announce the next item on the programme. He made a slight turn from the mike and said with a gruff but unfortunately clear enunciation: "That is the last time anyone will get me to talk into a damned little box."

After the first four years, responsibility for the school broadcasts was taken over from Victor Seary by Gerald J. Redmond, another recruit to the Department from Dalhousie University. Mr. Redmond had graduated in political science and joined the Department as Registrar in charge of teaching licences, which gave him charge of school broadcasts as well. He organized the broadcasts over the next thirteen years, during which time they grew in importance and prepared the ground for extension to the whole Maritime region.

Emphasis on Rural Schools

In the spring of 1937, the Nova Scotia Department of Education changed its policy in school broadcasting. Mr. Redmond found that the rural schools of the province were making greater use of the Department's broadcasts than the city schools, particularly the urban high schools. Dr. Munro and he therefore decided to shift the emphasis of the school broadcasts to rural and village schools. During May 1937 the Department conducted a month of experimental broadcasts on a daily basis for fifteen minutes each day, with the following programme schedule:

> *Monday*: science for grades 4, 5, and 6
> *Tuesday*: English for grades 7 and 8
> *Wednesday*: French for grade 7
> *Thursday*: history for grades 7 and 8
> *Friday*: history and civics for grade 9

This experiment was so successful that in the following October the old two-hour Friday period was transformed into a series of daily fifteen-minute broadcasts. These programmes were of a dual character. Each week three of the radio lessons were based on the prescribed course of study, while two others were of a supplementary nature and were presented on other days. The lessons based on the course of study were presented by selected teachers from the schools of the city of Halifax and served a double purpose. They were used by teachers in all sections of the province as model lessons to help them in their own class-room presentations. At the same time the broadcasts were made attractive to and comprehensible to the student audience in rural schools. But while the contents of the broadcasts were now based on the course of study, the material presented was by no means merely a repetition of class-room work. The radio lessons included a wealth of information not available to most teachers in rural and village communities. The supplementary programmes were designed for reception in all schools, urban as well as rural. The new orientation of the school broadcasts was well received by teachers in all sections of the province, who found them of increased assistance in their teaching work.

One feature of the new type of school broadcast was a weekly fifteen-minute French lesson. This series, originally begun by Professor C. H. Mercer, was continued by Mr. A. W. Cunningham, and subsequently taken over in 1940 by Mr. R. Burns Adams of St. Patrick's (Boys') School. Mr. Adams had had an earlier introduction to school broadcasting, for in December 1928 he was one of a group of high school students present in the studio to take part in the first French lesson given by Dr. C. H. Mercer.

About this time several amusing incidents were recorded in the Halifax studio. Early in 1937 an unintentional "cliff-hanger" in a dramatization of Canadian history produced one of the longest examples of suspense on record. The action of the drama had reached a point where the hero, tied to a stake and tortured, was menaced by an Indian chief brandishing a tomahawk. Just as the Indian chief declared to his victim "Now you die!" the station went off the air because of technical difficulties. The difficulties continued for two days, and it was not until the following Friday afternoon that the school audience was informed that the hero had escaped death after all!

Naturally, speakers giving school broadcasts were carefully instructed that their scripts must be timed to fill exactly the required period of fifteen minutes. School listeners in the spring of 1938 must have been greatly amused when a speaker on nature study suddenly stopped in mid-sentence at approximately the five-minute mark in a fifteen-minute talk, and was unable to continue. What the students did not know was that the speaker had numbered his pages on the reverse side of the paper, but had omitted the numeral 7. When he finished page 6 he shuffled through all of this sheets, failed to find page 7, and in complete confusion said "good afternoon" and faded off the air.

NEW BRUNSWICK AND PRINCE EDWARD ISLAND

The success of the Nova Scotian school broadcasts often prompted the question, When would school broadcasting be extended to cover the whole of the three Maritime provinces? From time to time individual schools in New Brunswick reported good reception and profitable utilization of the programmes from Halifax, but neither the New Brunswick nor the Prince Edward Island departments of education showed any sign of readiness to initiate school broadcasting on a province-wide basis. The only satisfactory alternative was to develop a co-operative presentation of the three provinces as a unit, an undertaking made easier by the fact that their courses of study and examination systems were very similar. What was required was some means of establishing a satisfactory network of stations to give adequate coverage throughout the three provinces, possible only through the good offices of the CBC.

The Contribution of the CBC

The CBC, conscious of its functions as a national public service broadcasting agency, was ready to co-operate with the three departments and give whatever help it could to create a Maritime school broadcast system. In fact, the CBC, both regionally and nationally, took the initiative in bringing the departments together to discuss how to overcome the obstacles facing them. At the time of the Royal Visit to Canada in June 1939, the CBC had opened a new high-powered regional transmitter, CBA, at Sackville, New Brunswick, which gave coverage over a considerable part of the Maritime region not hitherto reached. In February 1941, Donald W. Buchanan of the CBC Talks Department tried unsuccessfully to promote co-operation among the three provinces, but failed because New Brunswick did not yet feel ready to participate. By the following year, however, the position had changed in several respects. For example, during January and February 1942, the CBC had presented an experimental series of six national school broadcasts entitled "The Birth of Canadian Freedom" which greatly stimulated the interest of educators throughout the Dominion. Further, on April 10, 1942, the CBC called together in Toronto the first of a series of conferences among those concerned with school broadcasting in Canada as a whole. The conference was attended by, among others, representatives of the departments of education in the West, in Ontario, and in Nova Scotia. Mr. E. L. Bushnell, CBC General Programme Supervisor, presided and was supported by Mr. C. R. Delafield, CBC Supervisor of Institutional Broadcasts, and the writer, then CBC's Educational Adviser.

At this conference, G. J. Redmond, Nova Scotia's Director of School Broadcasts, reported the success which had attended the new provision of school broadcasts to rural schools in Nova Scotia:

At present five of the eight series of school broadcasts are radio lessons based closely on the prescribed Course of Studies (in English, French, Reading and History) for schools in the province. The remaining three series are of a supplementary character (covering Vocational Guidance,

Current Events and Citizenship). These lesson broadcasts do not consist merely of talks. The teacher at the microphone has to teach the lesson, just as if he were in the classroom—and usually with the aid of an actual class present in the studio. During the past year, there has been a course of Primary Reading aimed at students in Grade I. The studio class which participated in this consisted of 4 six year old and 2 seven year old children. There was also a Grade 7 French course, and a Grade 9 History course, and also two courses in English for Grades 10 and 11. The primary Reading Course has proved popular with mothers who had children of pre-school age at home; some of these have actually learned to read with the aid of the broadcasts. As regards the best time of day for school broadcasts, 10:00 A.M. is the ideal time in Nova Scotia, or alternately 2:30 in the afternoon. The courses of study for the provinces of New Brunswick and Prince Edward Island are similar to those of Nova Scotia, and the same school broadcasts would suit the needs of schools in all three provinces. The Nova Scotia school broadcasts are known as "The Maritime School Broadcasts", and are all presented from Halifax. Information about the programmes is published monthly in the Provincial Journal of Education, which circulates to all teachers. In addition, each teacher making use of the radio lessons is sent a supplementary bulletin once a month, setting forth project work, exercises to be done, etc. The duration of all school broadcasts in Nova Scotia is 15 minutes, except for one half-hour period of "Vocational Guidance" on Monday mornings. During the coming year, it is hoped to include a Current Events programme for Junior High School students, and a fifteen minute dramatization once a week on Citizenship. As regards the supplementary type of school broadcasts, there is plenty of room in Nova Scotia for [the inclusion of] national and international programmes.

In reply to the writer, who asked whether teaching broadcasts would not tend to displace the teacher in the class-room, Mr. Redmond asserted that there was no danger of supplanting the teacher. These lesson broadcasts provided model demonstrations of how teaching should be done, and were of particular service to young rural school teachers who had only had one year's training in the provincial normal school. There were then about 300 uncertificated teachers working in rural schools.

The CBC, for its part, had by now agreed with the Columbia Broadcasting System of the United States to make available in Canada, as needed by the educational authorities here, certain of the broadcasts of the CBS "School of the Americas." In addition, the conference in Toronto endorsed the desirability of continuing, on a regular basis, the provision of national school broadcasts with a common interest to Canada as a whole.

The first national series, "Heroes of Canada" (a full account of which is given in a later chapter), went on the air weekly from October

8, 1942, to March 19, 1943. These programmes, dramatizing famous figures in Canadian history, were contributed by the various provincial authorities. Nova Scotia was represented by two programmes, one of which, "Richard Uniacke—Dreamer of the Union," was produced in Halifax by Mr. Redmond from a script written by Dr. J. S. Martell, the Assistant Provincial Archivist. New Brunswick was represented by a programme on "Sir Brooke Watson—the Dick Whittington of the Maritimes" (December 11), and Prince Edward Island by a programme on "John Stewart of Mount Stewart" (January 22).

NEW BRUNSWICK AND P.E.I. JOIN IN

All these developments helped to pave the way for a meeting of the three Maritime departments of education, the CBC, and other Maritime educators and broadcasters. Two meetings were held in the first half of 1943. On March 27 representatives of the departments of education of Nova Scotia, New Brunswick, and Prince Edward Island met in the Brunswick Hotel at Moncton with representatives of eight private stations in the Maritimes, three teachers' societies, the Home and School Federation, and the CBC to consider planning a joint provision of school broadcasts in the Maritime region for 1943–44. Among those present were the Honourable C. H. Blakeny (Minister of Education, New Brunswick), Gerald Redmond (Nova Scotia Department of Education), George Young (CBC Maritime Region Director), H. M. Smith (CBC Regional Engineer), the writer (CBC Educational Adviser), Colonel K. S. Rogers (station CFCY Charlottetown), Major W. C. Borrett (CHNS Halifax), Stuart Neill (CFNB Fredericton), Stanley Chapman (CKNB Campbellton), F. L. Lynds (CKCW Moncton), T. F. Drummie (CHSJ Saint John), N. Nathanson (CJCB Sydney), L. Smith (CJLS Yarmouth), Miss Bernice MacNaughton (New Brunswick Teachers' Association), and L. A. DeWolfe (Nova Scotia Home and School Federation).

At this meeting it was agreed that the CBC would provide free time on the air over CBA and CBH, together with production facilities, on the understanding that the departments of education would, at their own expense, provide the programmes. Nova Scotia declared itself willing to continue to spend the sum already allocated to its own provincial school broadcasts ($500 a year) on the new Maritime series. The Honourable C. H. Blakeny promised to contribute programmes from New Brunswick, upon the understanding that the CBC would shortly appoint a programme organizer for the Maritime region who would give his full time to producing school broadcasts and travelling about the region to develop the work. The private stations offered to give free time on the air up to fifteen minutes every week-day, and it was agreed that these facilities, together with those offered by the CBC, would provide satisfactory coverage over the greater part of the region. Accordingly the meeting decided that in September 1943 the three provinces would co-operate to present daily school programmes over the Maritime network, and a committee was appointed to draw up a syllabus of subjects suitable to the courses of study in the three Maritime provinces.

Later in June this committee met at the Marshlands Inn at Sackville. The committee found that, to do an effective job gathering contributions from the three provinces and from Canadian national and American series, two fifteen-minute periods per school day would be required. These the private stations later agreed to carry.

FIRST MARITIME REGIONAL PROGRAMME

The first programme of Maritime school broadcasts was launched in the fall of 1943, and was carried over a total network of eleven Maritime radio stations. Details of the broadcasts were published in the Nova Scotia *Journal of Education* and elsewhere. The subjects were as follows:

Monday: 10:45–11:00 A.M., French for grade 7, given by R. Burns Adams of St. Patrick's (Boys') School, Halifax; 3:00–3:15 P.M., elementary science, grades 4–6, given by C. K. R. Allen of Queen Elizabeth High School, Halifax

Tuesday: 10:45–11:00 A.M., "Our Historic Past" (historical dramatizations from Halifax) for junior and senior high school; 3:00–3:15 P.M., vocational guidance for junior and senior high school, given by E. K. Ford, Halifax

Wednesday: 10:45–11:00 A.M., world geography for grade 7, given by W. Stuart MacFarlane, Principal of New Albert School, Saint John, N.B., and assisted by Travis Cushing, another school principal; 3:00–3:15 P.M., "New Horizons" (dramas of world history and geography) for junior high school, contributed by CBS School of the Americas

Thursday: 10:45–11:00 A.M., junior school music for grades 1–3, given by Miss Irene McQuillan, Assistant Supervisor of Music, Halifax; 3:00–3:15 P.M., "Tales from Far and Near" for junior high school, contributed by CBS School of the Americas

Friday: 10:45–11:00 A.M., national school broadcast (from Toronto); 3:00–3:15 P.M., music appreciation for junior high school, given by Professor Harold Hamer of Mount Allison University

The general purpose of the Maritime series was stated to be "to present programs which will provide new interests and appreciation and help in building desirable attitudes and ideals. The aim, while presenting a limited amount of factual material, is to supplement the work of the classroom teacher on the imaginative side." A leaflet giving particulars of the courses was published and distributed among teachers, and prizes were offered for the best scripts on Maritime history, submitted from schools in the three provinces.

On October 7, 1943, a special broadcast inaugurating the Maritime school broadcasts was given in which the education ministers of the three provinces took part. Premier A. S. MacMillan of Nova Scotia referred to school broadcasting enthusiastically: "Seated at his desk in the class-room, the child is brought by radio into close contact with life and experience. Here is equality of opportunity in education at work. Every boy and girl in every school—be that school in an urban, village or rural community—may have equal access to enriching influences through the presentation of school radio broadcasts." The Honourable C. H. Blakeny pointed out that school broadcasts were not intended to supplant the work of the teacher, but to lead the minds of boys and girls to search for new truths and to bring to them "a greater appreciation of the kind of world in which God intended we should dwell." Similarly, Premier J. W. Jones of Prince Edward Island said that "radio can supply the teacher with new ideas and subject matter which will prove instructive as well as entertaining, and will guide her in a general way in her programme of teaching."

During the ensuing winter, in pursuance of its promise given to the Honourable C. H. Blakeny, the CBC appointed Mr. Douglas B. Lusty, formerly a teacher and school music supervisor in Ontario, to become its School Broadcasts Organizer for the Maritime region. Mr. Lusty threw himself into the work with enthusiasm and tact. He succeeded in a short time in winning the confidence of the three Maritime departments of education and of the private stations in the area—a remarkable achievement for a newcomer from "Upper Canada." Lusty also showed himself to be a skilful publicist, particularly in the Maritime press. He possessed valuable talents for developing public relations among parents and teachers as well. In addition to promoting classroom interest in school broadcasts, he fostered the growth of a large adult (parent) audience which followed the programmes at home, and subsequently helped to mould public opinion in favour of the new medium.

During the next few years the Maritime school broadcasts steadily gained strength through the support of leading educators in the three provinces. Dr. Fletcher Peacock, who was appointed Director of Educational Services in New Brunswick, and later became Chief Superintendent of Education, was an unswerving and judicious supporter of school broadcasting, and regularly attended the meetings of the committee that planned the programmes, which met usually twice a year. Another firm supporter was Dr. Lloyd W. Shaw, who in 1944 left Carleton College, Ottawa, to become Superintendent (later Deputy Minister) of Education in Prince Edward Island. Although the Island did not feel able to contribute financially to the Maritime school broadcasts, Dr. Shaw himself (with the assistance of Mr. C. Ralph McLean of Charlottetown) personally contributed for many years numerous series of talks on agricultural and general science, which won warm appreciation in Maritime schools, and were later heard on request in Ontario and Quebec. Another contributor, of interesting and stimulating talks on English literature, was Dr. Albert Trueman, Superintendent of the School System, Saint John, who, after serving as President first of the University of Manitoba and later of the University of New Brunswick, then as Chairman of the National Film Board, was appointed Director of the Canada Council.

Among the most successful Maritime school broadcasts was the junior music programme directed by Miss Irene McQuillan (now Mrs. Alban Murphy) who is now Supervisor of Music in Halifax City Schools.

Miss McQuillan's junior music series has been on the air every year for the past nineteen years, and is still (1961) continuing. Originally, Miss McQuillan brought with her to the studio a trio of talented children, June Jollimore and John and Catherine Arab, whom she directed in a repertoire of songs suitable for elementary class-room singing. (John Arab later became a leading tenor with the Canadian Opera Company, while Catherine has continued for many years to sing contralto in the Armdale Chorus.) Each song that the trio sang was included in three successive broadcasts, and was thereafter constantly repeated until the class-room audience knew it by heart. Two-part singing was encouraged by the inclusion of simple rounds with incidental learning of the sol-fa syllables.

Miss McQuillan tells several interesting anecdotes of the early days of her music programmes. One of the most popular in her repertoire of songs was a melody which she picked up from listening to a group of Polish sailors singing while they played cards during a visit to Halifax harbour. Miss McQuillan persuaded the ship's officers to tell her the gist of the song, which enabled her to compose suitable English words to fit the melody. This song has ever since been used on the Maritime school broadcasts. The same song furnishes an apt illustration of the "contagious" effects of school music broadcasting. One day when visiting Evangeline, Digby, Miss McQuillan heard a class in a small school singing a Polish folk song which she identified as being the one she had introduced on the Maritime broadcasts. Knowing that this particular school possessed no radio and therefore could not receive her programme, she questioned the children and found out that one little girl had picked up the song by ear while listening to the radio at home during a bout of sickness, and had afterwards taught it to her class-mates at school well enough for them to sing it recognizably.

In 1944 at the suggestion of Douglas Lusty, Miss McQuillan undertook to present, with a choir drawn from the Halifax schools, a programme of Christmas carols to be heard over the national network on the Friday morning preceding Christmas Day. This broadcast proved so popular that it led to the institution, by the National Advisory Council on School Broadcasting, of a regular series of Christmas carol recitals given by school choirs, selected in rotation from every province in Canada.

FRENCH SCHOOL BROADCASTS IN NEW BRUNSWICK

For some time prior to 1954, the Society of Acadian Teachers and other French-speaking organizations in New Brunswick had been urging the Department of Education to make school broadcasts in their own language available to the French-speaking schools of the province. They pointed out that these schools now accounted for a slight majority of the school population of New Brunswick. However, it was not possible to meet their wishes because the French-speaking schools were widely scattered over different parts of the province, and there did not exist a sufficient number of French-language radio stations to assure adequate coverage for the proposed school broadcasts.

In 1953, however, the CBC opened a new key French-language radio station, CBAF Moncton, which gave renewed impetus to the proposed extension of school broadcasts. The teachers' representations were taken up by Dr. Gérard De Grace, at that time assistant to the Chief Superintendent of Education (now assistant to the Deputy Minister of Education) in the province, who showed great enthusiasm for the project. Dr. De Grace raised the question with Mr. Max Hickey, Director of the Audio-Visual Education Bureau of New Brunswick, and Mr. D. B. Lusty, CBC School Broadcasts Organizer for the Atlantic region. The result of their deliberations was a decision to provide school broadcasts in French on three days each week for the schools of New Brunswick, on the same basis as the English-language school broadcasts already provided by the Department of Education, jointly with the CBC.

The usual pattern of collaboration between broadcasters and educators was followed. The Department of Education assumed responsibility for the choice and content of the programmes, and for the selection and payment of the teachers who delivered the broadcasts. The CBC undertook to provide the studio facilities, time on the air, and network arrangements, on the same basis as prevailed generally in school broadcasting throughout Canada.

Accordingly, in October 1954, the first schedule of French school broadcasts was inaugurated, the Department presenting three series of fifteen-minute broadcasts weekly. On Tuesdays "Chantons Ensemble" was presented for grades 1–3. The aim of this series was to develop musical taste and add a finishing touch to its interpretation and presentation. The series was conducted by the Rev. Sister Marie-Lucienne, who directs the well-known choir of Notre Dame d'Acadie. The programmes consisted largely of Acadian folk songs, in which the listening school pupils were invited to join. On Wednesdays "Parlons Mieux le Français" for grades 4–6 was given, its aim being the study of French pronunciation and the use of the correct style of speech. Mlle Léonie Boudreau gave the broadcasts, often using a French pupil at the microphone, and inviting her school listeners to participate in the repetition of words and phrases. On Thursdays "Parlons Anglais" was presented for grades 6–9. It was a course of English, regarded as a second language, and was aimed at children whose teachers might have some difficulty in making their lessons in English effective. The broadcaster was M. Hyacinthe Leblanc, the school principal at Edmundston. The first year's course was entitled "Une Journée Bien Remplie." The French-language stations, CBAF Moncton, CHNC New Carlisle, and CJEM Edmundston, carried the broadcasts from 10:00–10:15 A.M., AST.

The broadcasting of these programmes over CHNC New Carlisle meant that they could be heard in some schools in Quebec, as well as in New Brunswick.

The three series were well received in the French-speaking schools, both by pupils and by teachers. According to Dr. De Grace "they have supplied a real educational service. The children sing more and better than they did previously. The speech habits of the young class-room listeners indicate real progress, and the teaching methods of many of their instructors have shown considerable improvement." The first two courses became a permanent feature of the schedule. The third course, however, did not succeed so well in reaching those pupils who had most need of it. Accordingly from 1959 onwards it was replaced by another course of broadcasts, "Les Belles Lettres," which aimed at promoting the appreciation of good children's books. The new course proved at once to have value, as was shown particularly by the many requests, directed to the central circulating library of the Department of Education, for the books discussed.

III. British Columbia

As early as 1927 station CNRV Vancouver co-operated with the Vancouver School Board in presenting on Friday evenings an interesting series of educational programmes by children for children. Mr. G. S. Gordon, Inspector of Schools for Vancouver, with the principals and teachers of twenty-four city schools, prepared the series, to which one room from each school in turn contributed a 60-minute programme of songs, educational games, and helpful lessons. The musical parts of the programmes were directed by Miss A. Roberts, Assistant Supervisor of Music for Vancouver. Though primarily intended for children, these programmes attracted also a large adult audience. But the most marked response came from country teachers who in many instances arranged for groups of children to gather in a home where there was good radio reception, to listen and learn. Some children came as far as fifteen miles to attend, while in one locality children came in rowboats across deep channels and through narrow fjords. Such was the lure of radio in those early days! CNRV also received many letters from teachers who wanted further instructions on how to follow up the lesson-broadcasts. At the wish of a meeting of Vancouver school principals, the broadcasts were continued in 1928–29.

The beautiful Okanagan Valley, however, in the heart of British Columbia, may justly claim to be considered the nursery of school broadcasting in the province. Not only was the first experiment conducted there by the teachers of the district with the co-operation of the radio station at Kelowna; but also many of those who subsequently played a leading part in the development of the provincial and national systems of school broadcasts—such as Dr. Ira Dilworth (Pacific Regional Director, CBC), Mr. Kenneth Caple and Mr. Philip J. Kitley (respectively first and second Directors of School Broadcasts of the British Columbia Department of Education), and Mr. Monteith Fotheringham (first Assistant Director of School Broadcasts)—either lived or taught in the Valley at the outset of their careers.

In 1936 the Okanagan Valley Teachers' Association, under the presidency of Mr. Kenneth Caple, then Principal of Summerland High School, found itself in the happy position of possessing a surplus in its funds. The Association decided to use this money for a short experiment with the use of radio in the class-room. The experiment was made possible by the co-operation of Mr. James Browne, the owner and manager of CKOV Kelowna, who provided without charge the time for the broadcasts. The first series consisted of six programmes in music appreciation conducted by Mr. F. T. Marriage, then Principal of a Kelowna elementary school, and Mr. C. S. Mossop, also of Kelowna (now Supervisor of School Music in Calgary). The programmes aroused considerable interest in schools in the Okanagan Valley. Later the series was continued as a weekly series of evening music programmes over CKOV.

In due course a report of the Okanagan Valley experiment was made to the British Columbia Teachers' Federation, of which Dr. Ira Dilworth (then Principal of Victoria High School) was President, and Mr. Harry Charlesworth, Secretary. The report was also forwarded through Inspector E. J. Frederickson, Principal of Kelowna Junior High School, to Dr. S. J. Willis, the Provincial Superintendent of Schools. Dr. Willis was sufficiently impressed to feel that the possibilities of radio in schools ought to be further explored.

APPOINTMENT OF SCHOOL RADIO COMMITTEE

Dr. Willis decided to appoint a committee of enquiry into the matter, under the chairmanship of Mr. Robert England, Head of the Extension Department of the University of British Columbia. This committee held one preliminary meeting at the Georgia Hotel, Vancouver, on May 7, 1937, but no further progress took place for the next three months. In the meantime Mr. England relinquished the chairmanship, whereupon Dr. Willis requested Mr. A. Sullivan, Inspector of Schools in Victoria, to reconvene the committee. The first formal meeting of the new Radio Committee was held on August 23, with Mr. Sullivan in the chair and Mr. A. R. Lord, Principal of the Normal School, acting as Secretary. Mr. R. H. Bennett, of John Olive High School in Vancouver, was also present. From the beginning Mr. Lord interested himself keenly in the proposed experiment, which he believed would have value in the training of teachers. He soon became the driving force behind the Committee, and devoted himself with energy and persistence to its work.

At this first meeting, the discussion centred around the feasibility of conducting a formal experiment in school broadcasting, with programmes twice a week for eight or ten weeks. The Committee decided to seek information on four important questions: the value of radio in teaching; the most useful types of programmes; the most suitable length of programmes, and the number of receiving sets available in the schools of the province. A circular letter covering these points was sent out to inspectors of schools, and the replies indicated that, at that time, only 26 schools possessed receivers; that two-thirds of the schools in the province had no electrical connections and would therefore require battery sets; and that a large area of the province could not receive any CBC programmes during daylight hours.

By the Committee's second meeting, on October 13, its membership had been enlarged by the addition of Mr. Harry Charlesworth, Secretary of the British Columbia Teachers' Federation, and Mr. William Morgan. The Committee proceeded to lay down certain general conclusions, drawn from its preliminary investigations. These were as follows: Nothing was to be gained by providing service in any field which was being efficiently developed by the teachers themselves. School broadcasts should be limited to those subjects in which teachers stood in greatest need of assistance—music appreciation, art appreciation, social studies, elementary science, and literature. The service should be developed mainly as an aid to rural schools where the need of outside help was greatest. Finally, to ensure success, a high degree of showmanship in the broadcasts would be necessary. For this reason they should, as far as practicable, be prepared and presented by professional radio people.

The next problem was the gaining of sufficient financial backing from the Department to put on a series of experimental programmes, to be received by approximately sixty schools in the province. The Committee estimated its requirements at $1000, and applied to the Minister of Education, the Honourable G. M. Weir, for this amount. Early in 1938 the Committee heard from Dr. Willis that a grant of

$500 would be made available for a ten-week series of experimental school broadcasts.

Then began a period of intensive preparation. The Committee met almost every week till the summer. Its first step was to apply to the CBC for help in the presentation and distribution of the broadcasts. Mr. Gladstone Murray, CBC's General Manager, was highly sympathetic to the British Columbia project. He instructed two of his staff, Mr. Jack Radford, the manager of station CBR Vancouver, and Mr. William Ward, to attend the committee meetings and give whatever advice and help might be needed. From the outset it was accepted policy that the CBC representatives would concern themselves primarily with the form of presentation, while the educators would completely control the planning and content of the proposed broadcasts.

Three courses were planned, with a special eye towards helping teachers and pupils in smaller rural elementary schools. The subjects selected for the experiment were as follows: "Musical Pathways"—ten programmes in music appreciation, prepared by a Vancouver Normal School instructor, Miss Mildred McManus, which included practice in the singing of songs familiar to the schools, with participation from the class-room audience; "Elementary Science"—five programmes dealing with water, weather, flying, birds and fishes, and so on, prepared by a science teacher, Mr. J. W. B. Shore; and "Social Studies"—five programmes dealing with Western history, including the North-West Passage, the fur trade, the gold rush, and so forth, prepared in dramatized form by Mr. Ward.

The Secretary of the Committee, Mr. Lord, undertook to examine and revise the draft scripts for all programmes.

For its part, the CBC allocated to the broadcasts the period 9:30 to 10:00 A.M. PST, beginning on March 21, 1938. The CBC also arranged for the programmes to be carried over CBR Vancouver and four privately owned stations, CHWK Chilliwack, CFJC Kamloops, CKOV Kelowna, and CJAT Trail.

Mimeographed outlines of the programmes were mailed out, before the series went on the air, to nearly 300 teachers in the areas served by the network. However, as soon as the broadcasts started, complaints of unsatisfactory reception came in from many rural schools that desired, but had not been able, to listen to the broadcasts. It became clear that the trouble was largely due to inadequate radio coverage in the interior of the province. The deficiency was made the subject of questions in Parliament, and was also discussed fully with Mr. Gladstone Murray. In due course it led to the erection by the CBC of repeater stations in the interior of British Columbia.

APPLICATION FOR A CARNEGIE GRANT

By the time that the experimental series was nearly over, the Radio Committee had begun to consider what should succeed it. Reports coming in from teachers indicated that the programmes had been successful, and that a regular year-round programme of school broadcasts was desired. On May 1, accordingly, the Committee adopted recommendations favouring a continuation and enlargement of the programmes in ten-week units during the next school year, with separate series for each grade level including high schools. A request was forwarded to the Minister for a grant of $3500 to cover the cost during 1939. Two weeks later a negative reply was received. For reasons of economy, no more money for school broadcasting could be included in the departmental estimates. The Committee did not accept the refusal as final, but sent a deputation to the Minister to press its claim. It could point to the fact that the school broadcasts had gained the enthusiastic support of the provincial Teachers' Federation and of the Parent-Teacher Federation, which on June 10 passed a resolution particularly favouring the "Musical Pathways" series. Also, the Inspector from the Peace River district was asking for help to equip the schools in his district with receivers; and station CFGP Grande Prairie was offering to carry the provincial school programmes by transcription. These representations had their effect. By the end of June, the Department of Education modified its attitude to the extent of granting the Committee $2500, to last until March 31, 1939.

The Committee felt able to proceed with its plans for the fall and winter of 1938–39. However, during the temporary financial uncertainty, another source of assistance was explored. On June 11, Mr. Lord wrote to the Carnegie New York Trustees, asking for a grant of $4000 to further the Committee's work. Five months later, in November, the Trustees replied favourably, granting the Committee $3000 for further experiments. The additional source of revenue made it possible to plan the school radio work for a two-year period; also to pay modest fees (approximately $15 a programme) to each script-writer employed.

In September, the Committee gained additional strength by the appointment of a distinguished new member, Dr. Ira Dilworth. In 1934 Dr. Dilworth had left Victoria to join the staff of the University of British Columbia, first as Associate Professor and subsequently as Professor of English Literature. In 1938 he took a leave of absence from the University in order to accept an invitation to become Mr. Gladstone Murray's representative in the British Columbia region, as well as manager of station CBR Vancouver. Two years later he resigned from the University and joined the CBC permanently. In his new position he became the CBC representative on the provincial School Radio Committee. Dr. Dilworth's wide experience and accomplishments in education, music, and literature made him a powerful force, not only in the West, but throughout Canada.

During the summer of 1938, Miss Margery Agnew, Vice-Principal of Vancouver Technical High School, announced the formation of the Sir Ernest MacMillan Fine Arts Clubs in many schools of the province. She also proposed to the Radio Committee that these clubs should be integrated with the music appreciation broadcasts. The Committee found itself unable to adopt this suggestion, but invited Sir Ernest MacMillan himself to deliver, in Vancouver, the initial programme in its new music appreciation series. Sir Ernest gladly accepted, but before he could give his broadcast, the Musicians' Protective Association intervened and forbade him to proceed. This was the first, but by no means the only, occasion in the history of school broadcasting when the development of school music through radio had to meet the problem of conforming to union regulations.

"BRITISH COLUMBIA RADIO SCHOOL"

Early in September, Mr. Lord announced the Committee's plans for what was now termed "The British Columbia Radio School" for the fall and winter of 1938–39. The broadcasts were to be given on five days a week (Monday to Friday) from 9:30 to 10:00 A.M. from October to March. The subjects were the following:

Monday: social studies for grades 4–6, guidance before Christmas, and history after Christmas
Tuesday: junior music for grades 1–4, "Mother Goose"
Wednesday: elementary science for grades 7–9
Thursday: senior music for grades 5–8, "Musical Pathways"
Friday: high school hour for grades 9–12, "Music and Poetry"

The programmes were heard live over the same stations as previously, and by delayed transcription over CFPR Prince Rupert and

CFGP Grande Prairie. The inclusion of the latter station meant that some Alberta schools would be listening to British Columbia school broadcasts. Accordingly, recordings of sample broadcasts were sent to Dr. G. F. MacNally, Alberta's Deputy Minister of Education, and met with his approval. This was the beginning of co-operation between the two provinces, which later expanded, with the inclusion of Manitoba and Saskatchewan, into a four-province operation.

Even at this date, some teachers raised their voices against the early morning period at which the school broadcasts were being scheduled. But because the CBC could not yet offer any suitable alternative network time, and the teachers could not agree what hour would best suit them, no change was made.

There were criticisms, too, of the programmes themselves. According to the departmental circular,

A number of teachers in schools in the larger cities report that programmes are too simple for certain grades and too difficult for others, or that they do not follow the [Provincial] Programme of Studies. Such criticisms are quite correct, but an immediate reply does not seem to be possible. The limited extent of available radio time makes it necessary to combine grades or to omit several of them. Further, all programmes are intended primarily for the smaller rural schools. It is possible that, later on, separate broadcasts for rural and for urban schools will be necessary. . . . But in the meantime, rural schools must receive first consideration. It is impossible to *follow* a Programme of Studies by means of radio. The sole purpose of broadcasts is to create a background or to add to what the teacher is able to do—but not to replace the teacher.

Another matter causing concern to Mr. Lord and his Committee was the payment by schools for licences to possess receivers. Early in September the Department of Education took legal opinion on this matter, and subsequently approached the Ministry of Transport in Ottawa. The result was a ruling that any public (but not private) school could get a free receiving licence provided application on its behalf was made by its Department of Education. The decision was welcomed in all provinces, and undoubtedly made it easier for schools to listen to the broadcasts. By the end of 1938, over 400 schools in British Columbia had taken advantage of the licence privilege, and had acquired receivers.

Some difficulty was experienced during 1938, through the wish of one of the private stations forming part of the provincial school radio network, CKOV Kelowna, to discontinue carrying the broadcasts, in order to sell the time commercially. After some negotiation with and pressure from the CBC, the station resumed giving coverage to the school programmes. However, the tendency of individual private stations to drop off the school network made itself felt, in varying degrees, in British Columbia and later in other parts of the country.

Early in 1939 Mr. Lord sent an outline of the progress made to date in school broadcasts in British Columbia to Dr. E. A. Corbett, the Director of the Canadian Association for Adult Education, who had been commissioned by Mr. Gladstone Murray to make a report on school broadcasting in Canada as a whole. Mr. Lord's outline contained the following interesting observations.

[*Suitability of Subjects*] Music comes easily first in such a list, followed by Art and Literature, each treated from the appreciative point of view. History and Geography, either separately or as Social Studies should be presented solely with the idea of creating a background for the regular classroom activities. Elementary Science, Health and Safety Education are recent additions to the curriculum and, for that reason, probably should be included. Guidance is also a possibility. . . .

[*Urban and Rural Schools*] We are continually confronted with a situation where a programme has proved very acceptable to a graded urban school, but was "over the heads of the pupils" in an ungraded rural school. More and more we are forced to conclude that the only satisfactory solution lies in two distinct sets of programmes. Perhaps each fair sized city (50,000 and over) will have its own broadcasting station. For the present, all broadcasts must be designed for the school where the need is greatest. . . .

[*Showmanship*] Lectures and talks, or lengthy narrations, are an even greater waste of time on the air than in the classroom. Dramatizations are effective, but can be easily overdone both in frequency and in exaggerated effects. The general speed must be slowed down, and greater attention paid to language, articulation, etc. . . .

[*Canadianism*] Radio is making better Canadians. One illustration will suffice. In the Peace River Block there are 55 schools, and at least as many nationalities are represented among the pupils attending them. It is much the most isolated area of commensurate size in British Columbia. Agriculture is the sole industry, and the resultant economic situation is acute. Living conditions are primitive, and knowledge of the outside world is limited, for newspapers, magazines, telephones, etc., are beyond the reach of most of the settlers. There are radios in at least 54 out of the 55 schools. At 1:00 P.M. daily there is a 10-minute news broadcast. It serves two purposes; pupils write down the chief items, which they take home for the information of the family, and these same items form the basis for a lesson in Social Studies. Can one conceive of any way in which Geography can be made more real?

Mr. Lord also discussed, in his outline, the responsibility of the CBC towards school broadcasts. He considered that the CBC should formally take responsibility for providing coverage wherever schools existed by extending its network to areas (such as the Cariboo country) which could not receive any Canadian station. Also the CBC should accept joint responsibility with the Department of Education for the preparation and production of all programmes. He agreed that the costs of these ought to be shared between the two parties, on the following basis: "(a) The Department of Education to determine the nature of all programmes, to exercise supervision over all scripts, language, etc., and to meet the costs of scripts, production and all literature to schools. (b) The CBC to be responsible for the technical phases of all productions and to provide the services of their stations without remuneration."

During the school year 1939–40, the programme of school broadcasts was continued on the same basis as previously, except that no provision was made for high school broadcasts, partly because of Dr. Dilworth's absence in Ottawa and partly because of the limited opportunities for listening available to high school students.

APPOINTMENT OF CAPLE AS DIRECTOR

By now, the task of organizing the programmes and supervising the scripts had become too great to be carried on solely by members of the Radio Committee, each of whom had his own full-time job. The need was evident for the appointment of a full-time officer to take charge of the work. The Department of Education, for its part, was ready to appoint such an official, but not to provide the full sum of money necessary for the post. Discussion, accordingly, took place between departmental officials, Dr. Dilworth, and Major Gladstone Murray, regarding the possibility of the CBC sharing the financial burden involved. Recognizing the importance of the school broadcasting work, Major Murray agreed that the Corporation should contribute one-half of the salary of the new official, and provide him with an office at the CBC's Vancouver headquarters. In this way, the officer would work in immediate propinquity to the broadcasting studios, and be able to share in the actual work of production.

Under these circumstances, the Department announced on November 6, 1939, its readiness to appoint a Director of School Broadcasts for the province. While the Committee was considering whom it might recommend to fill the post, further good news came to hand. On January 11, 1940, a promise was received from the Carnegie Trustees to make an additional grant of $2500 towards the salary of the proposed Director. Subsequently, a third grant of $1500 was made for the same purpose.

Two weeks later, at its eighteenth meeting, the Committee, on the advice of Dr. Dilworth, recommended the appointment of Mr. Kenneth Caple. Mr. Caple had been Principal of the Summerland High School in the Okanagan for ten years and had come to the attention of Dr. Dilworth and others as an "experimenter" with vision and initiative. From 1938 to 1940, as Director of the Dominion–Provincial Youth Training Project, carried out under the general supervision of the Department of Extension of the University of British Columbia, he established "folk schools" in the remote rural areas of British Columbia. Caple brought to the work on school broadcasting a broad background and varied experience in education, supported by much personal charm and enthusiasm for the work. During the next four years he was the moving force in the successful development and rapid extension of the school broadcasting work.

INNOVATIONS IN SCHOOL BROADCASTS

During the spring of 1940 Mr. Caple made an extensive tour of radio stations in the United States to acquaint himself with the progress of school broadcasting in that country. He then settled down to work on the programme schedule for the coming school year. Several important innovations were made. Separate schedules were planned for the autumn and spring terms. The length of most programmes for elementary schools was cut to fifteen minutes. (The Radio Committee had earlier arrived at the conclusion that twenty minutes was the "optimum" length for a school broadcast.) Thus more subjects could be covered in the year. Finally, each term a mimeographed 32-page teachers' manual was issued to teachers, giving full details about the programmes, as well as advice to teachers on receiving equipment and utilization of the broadcasts.

The introduction to the manual clearly expresses the general philosophy of school broadcasts in the province. "The purpose of school broadcasts is not to teach a curriculum, but to enrich it by bringing the world outside into the classroom. The broadcasts are not planned to supply specific factual material for special grade levels, but are planned to supply supplementary integrated materials suitable for

several grades. The aim of the in-school broadcast is to present programmes which will help to give new interests and appreciations, and
to build desirable attitudes and ideals." The same introduction stresses
the important part teachers have to play in class-room radio. "The
radio can be successful in schools *only* with the unceasing aid and
criticism of the class-room teacher and the pupils who listen."

TABLE I

PROGRAMME SCHEDULE FOR 1940–41

	Autumn term 1940		Spring term 1941	
	subject	grades	subject	grades
Monday	language arts	4–9	"Art on the Air"	3–8
			"Your Visitor Today"	All
Tuesday	junior music	1–2	junior music	1–2
		3–4		3–4
Wednesday	social studies	4–7	"Road to Democracy"	5–12
	newscast	4–12	newscast	6–12
Thursday	senior music	5–9	senior music	5–9
Friday	health	All	"Our Living World"	
			(science interviews)	5–9
			language arts	5–9

The programme schedule for 1940–41 was as given in Table I. The
opening broadcast on October 1 was introduced by the Minister of
Education, Hon. Dr. G. M. Weir. The network of stations carrying the
school broadcasts was that year increased to nine, two of which used
delayed transcriptions.

During the next two years many fresh programme ideas were introduced by Mr. Caple, and made the subject of experiment. Notable
among these were "The Mighty Fraser," a series of historical dramatizations, in which the river was the narrator, telling of the adventures
which it had seen on its course. The children in the class-room constructed large maps and filled them in as the story unfolded. This
proved so popular it was repeated in successive years. Highly popular,
too, was an elementary (grades 5–9) science series entitled "The
World's Great Wonderers," dealing with the lives of scientists. Each
programme was introduced by broadcasting the sound of a baby cry-
ing. Subsequently the cry was identified as that of the child who
developed into Galileo, Newton, Archimedes, or some other famous
scientific discoverer.

Another experiment, of a "fantasy" type, was a series "From Marco
Polo to Me" (grades 3–6), dealing with the history of transportation.

An imaginary character, Charlie Chickinick, was created, who was taken by the Spirit of Progress back from the present day to the Middle Ages, where he experienced various adventures in company with different great explorers.

Another series, produced by Mr. John Barnes, broadcast in the fall of 1943 and repeated in 1944, was "Working Together in Tukwilla Valley," the story of an imaginary community in the province where the teachers, children, and parents mobilized all available resources to ensure the best in modern education. Many listening schools undertook local projects of their own, modelled on these broadcasts. In one of these projects, a farmer gave the local school committee a barn, which they transported down the road and attached to the schoolhouse to make a gymnasium. After the broadcast, the National Film Board made a film of "Tukwilla Valley," which was widely shown throughout the West.

All these new programme ideas helped to stimulate interest in the school broadcasts. By the winter of 1941–42 the number of schools in the province equipped with radio had grown to 527, including a few (such as at Nanaimo on Vancouver Island) with central public address systems enabling every class-room to tune in to the broadcasts.

During the same year, in response to requests from many teachers, the time of day at which the school programmes were broadcast was changed from 9:30–10:00 A.M. to 2:00–2:30 P.M. This afternoon period became a permanent arrangement.

In 1944–45 the chief programme innovation was the staging of a series of high school forums, five of which were broadcast and afforded valuable opportunities for the participation of high school students. However, as Mr. Caple reported, "while these programmes are used enthusiastically by some secondary schools, it is doubtful whether the series as a whole is listened to as consistently as other programmes."

PROGRAMMES EXTENDED BEYOND THE PROVINCE

For some years previously, the school broadcasts in British Columbia had been of interest to many schools outside the province. They were carried, by arrangement, over the local station at Bellingham, across the American border, and were also heard in many Alberta schools. In 1939, after the CBC had opened its new high-power Prairie regional transmitter at Watrous, Saskatchewan (CBK), Mr. Gladstone Murray agreed that two of the British Columbia school series (music appreciation on Tuesdays and science on Thursdays) should as an experiment be carried by the transmitter to the whole of Western Canada. This was done in 1940–41 and gave a great fillip to the idea of establishing a co-operative series of broadcasts covering the four Western provinces. Two such series, "Highways to Adventure" from Vancouver and music from Winnipeg, were presented in 1941–42 (see pp. 54–5). Dr. Newland of Alberta and Mr. Caple of British Columbia were the moving forces in this co-operation, which took definite shape at a meeting at Banff, Alberta, on August 3, 1943, when the Western Regional Committee was established. A four-province series of intermediate and senior music went on the air in the following year, 1944–45.

In the spring of 1944, when the first meeting of the National Advisory Council on School Broadcasting was held in Toronto, Mr. Caple attended as the representative of British Columbia. As convener of the Council's programme committee, he introduced a report containing many valuable suggestions for future series of national school broadcasts. Shortly after the meeting, however, he accepted an invitation from Dr. Dilworth to join the permanent staff of the CBC, as Programme Director for the British Columbia region. He had therefore to relinquish his special responsibility for the school broadcasts, although he continued in his new capacity to take a keen interest in their progress.

APPOINTMENT OF MR. P. J. KITLEY

To succeed Mr. Caple the Department of Education, in consultation with the CBC, appointed Mr. Philip J. Kitley, an elementary and junior high school teacher in Kelowna. Mr. Kitley had come to the attention of the district high school inspector when during 1943–44 he conducted a successful series of music appreciation broadcasts over CKOV Kelowna. In June 1944 Mr. Kitley was in Vancouver in connection with marking the Departmental Examination papers when he received the new appointment. Like Mr. Caple, he brought to the school broadcasting work a mind fertile in new ideas, and a special interest in the improvement of class-room receiving equipment and in the use of electronic devices in education.

Mr. Kitley's first step was to turn the provincial school broadcasts teachers' bulletin from a mimeographed into a printed and illustrated booklet. It appeared for the first time in connection with the 1944–45 programme schedule. His keen interest in music soon led to a strengthening of the music appreciation features in the school programmes,

which was reflected in the 1945–46 schedule. The programmes for that
year were as follows:

Monday: "Over the Counter" for grades 4–8 (social study programmes on
the history and production of everyday commodities); "Magic Hinges"
for grades 1–3 and 4–7 (stories for juniors); elementary science for
grades 5–9
Tuesday: Music for juniors for grades 1–4
Wednesday: "Canadian Personalities," and other short series for high school
for grades 7–12
Thursday: intermediate and senior music for grades 5–9

For the Thursday series, extensive use was made of phonograph
records for music appreciation. Mr. Kitley himself became the organizer and narrator for the series "Listening Is Fun," whose theme was
"appreciation of great instrumental artists," and which he conducted
regularly every year from 1945 on. While conducting this programme,
Mr. Kitley had various experiences of the problems involved in music
broadcasts. On one occasion he put on the air a programme including
"live" musicians without the prior permission of the Musicians' Union,
producing an uproar that can well be imagined! On another occasion
when the broadcast dealt with pianos, Mr. Kitley innocently remarked
that "Steinway" was synonymous with "grand piano." His telephone
resounded with complaints from numerous other piano manufacturers, to mollify whom he had to make a public apology on his next
broadcast. Occasionally he had "audience trouble" too. Mr. Kitley
sometimes allowed parents to bring children into the control room to
watch production. This had to be stopped after one child, placed on
the control room panel to see the show better, pulled all the panel
plugs and took the programme off the air!

In 1947–48 a new series "Pictures on the Air" (art for grades 4–8)
met with great popularity in the schools. The series was similar to
Manitoba's series, "It's Fun to Draw," which started at the same time.
In both cases the broadcasts were planned to encourage free expression in art on the part of listening students. The British Columbia
programmes, however, were more tutorial in style than the Manitoba
programmes. Mr. Kitley thought it desirable to give weaker students
of art some definite hints on subject and technique, without restricting
the freedom of the more capable students. "Pictures on the Air" won a
permanent place in the provincial schedule.

The same year saw the start of "Ecoutez!" a regular series of programmes designed to enliven French studies at the senior high school
level. The aim of the broadcasts was to give enrichment, through

dramatizations and conversations, and to stimulate the motive to study and speak French. Music and variety were often featured in the broadcasts.

OVERCOMING TECHNICAL HANDICAPS

The progress of the school broadcasts at this period was considerably hampered by lack of technical knowledge and poor receiving equipment in rural schools. Mr. Kitley made an effort to overcome this handicap by thoroughly investigating the problem of radio reception in schools. He enquired into the possibility of using surplus war assets to relieve the radio shortage. He drew up a guide for teachers and students on technical matters. At his suggestion the departmental school building manual was expanded to include information which he had prepared on school central sound receiving systems (covering loud speakers, sound recorders, phonographs, and so on). Also the provincial Research Council undertook to examine and judge any specifications for central sound systems submitted to it by school suppliers.

By 1947, the overcoming of these technical handicaps, the increased coverage resulting from the establishment by the CBC of new repeater stations in the interior of the province, and the activities of Mr. Kitley in addressing meetings of teachers and parent–teacher groups, as well as visiting individual schools, were beginning to show favourable results. The number of listening schools in the province rose to over 1100.

In 1946 the long-standing arrangements between the CBC and the Department of Education, whereby the salary of the Director of School Broadcasts was shared equally between them, came to an end. Henceforth the Director became an officer appointed and paid solely by the Department of Education. However, the CBC continued to provide an office for his work conveniently adjacent to the studios and production offices of the CBC. In 1947 the Department decided to encourage the further development of the school broadcast work by appointing an Assistant Director of School Broadcasting, and Mr. Monteith Fotheringham, a Kelowna teacher, was chosen to fill this position.

In 1950 the Committee for Radio in Schools, which had played such an important role in guiding the school broadcasts in British Columbia through their initial and formative periods, was dissolved, with the approval of the Deputy Minister.

IV. The Prairie Provinces

ALBERTA

School broadcasting in Alberta traces its origins indirectly to an innocent piece of deception practised at the University of Alberta in 1927. Five years previously, Mr. H. P. Brown, who at that time was in charge of the visual-aid department of the University, began to interest himself in promoting the educational possibilities of radio. He succeeded in enlisting the support of the Director of Extension, Professor A. E. Ottewell, but no practical progress could be made in the absence of any available funds to implement Mr. Brown's favourite project, the establishment of an educational radio station. In 1925 Mr. Brown began carrying a radio programme over the commercial station CJCA and in 1926 he constructed a small studio (later rebuilt) from which music and lectures could go to CJCA by remote control. Thus, the principle of a radio programme had been established but the time available on CJCA was insufficient and it was necessary for the University to establish its own station.

In 1927, the University budget included the sum of $7,000 to hire an additional lecturer in the Extension Department. What followed is quoted from an article by Mrs. Dorothy Dahlgren in *Alberta Calls* (staff magazine of the Alberta Government Telephones) in its January-February issue of 1960:

Several months passed during which nobody noticed that the new lecturer never did arrive on the scene. Nobody noticed that a number of electrical engineering students were suspiciously busy in their spare time building (with the aid of Mr. W. Grant, then operating Radio Station CFCN, Calgary), a radio transmitter and antenna. When it was finished, the Department of Extension bought two windmill towers 75 feet high, added some old iron poles to make them 100 feet high, and attached the antenna to them. The transmitter was installed in some small shack behind the Arts Building. Then they proceeded to fix up the studios in what is now the University Power House. With all this done, the new little station was ready for business.

Mr. Brown estimates that the total cost of the station controls and studios was approximately $7,500.

Start of Station CKUA

There was trouble over procuring a licence from the Department of Transport and, of course, over the use of the University grant. However, these and other difficulties were one by one overcome. By purchasing one of three small stations already operating in Edmonton and changing its call letters to CKUA, the University of Alberta radio station at last came into being, and in November 1927 made its debut to the public of Alberta. CKUA began operating with one paid staff member, and its activities were at first decidedly limited. Naturally its operation came under the control of the University Extension Department which used it mainly during the first ten years for adult educational purposes. However, on May 23, 1929, the first experimental school broadcast in Alberta was produced, and it was heard in a number of schools equipped with receivers on loan by local radio dealers. In 1932 French lessons were broadcast, and used by many high schools in and around Edmonton. For example, a report from Vegreville High School stated that well over half of all students studying French were taking lessons from CKUA. Then in September 1936 the schools of Edmonton were unable to open because of an epidemic. To meet the emergency, lessons were broadcast from CKUA to the children in their homes.

Conference of Educators

It was not until 1937, however, that, through the efforts of Mr. Donald Cameron (later Senator Cameron) who was then Director of Extension, the University authorities approached the Supervisor of Schools for the Province, Dr. H. C. Newland. Dr. Newland, though himself a progressive and vigorous educator, did not yet see any great future for school radio broadcasting. However, he was impressed by the fact that during the first three months of 1937 station CKUA had broadcast a series of teachers' forum broadcasts which had been well received and had stimulated teachers' interest in the use of radio to help them in their class-room work. Accordingly, he undertook to call a conference on April 24 of educators interested in broadcasting. Present at this conference, over which he presided, were Dr. G. F. McNally, Deputy Minister of Education; Mr. Donald Cameron, Director of the Department of Extension, University of Alberta; Miss Sheila Marryat, Director of station CKUA; Dr. M. E. Lazerte, Director of the School of Education, University of Alberta; Dr. G. S. Lord, Principal of the Edmonton Normal School; Dr. John R. Tuck, President of the Edmonton Education Society; and Mr. H. C. Clark, representing the Alberta Teachers' Association.

The conference, after favouring a continuance of the teachers' forum broadcasts, proceeded to take up the more important subject of radio broadcasting to schools. This was a matter, it was held, for which the Department of Education itself must assume responsibility. Therefore, the Minister of Education was asked to set up a departmental radio committee, composed of representatives of the Alberta Teachers' Association, the Alberta School Trustees' Association, the School of Education of the University of Alberta, the normal schools, the University of Alberta (Department of Extension and station CKUA), and the Alberta Department of Education. This Committee lost no time in meeting on May 15, 1937.

The Lethbridge Experiment

It arranged for an experimental regional school broadcast to be conducted in the Lethbridge area in the winter of 1937–38 on the following basis: (a) the programmes to be of a kind especially helpful to underprivileged schools; (b) each school broadcast to be not longer than fifteen minutes and, on an average, schools to receive not more than two programmes per day; (c) the programmes to be aimed at grades 1 and 2 of the elementary school, and also the intermediate school; (d) the subjects to be as follows: music and music appreciation, social studies, science, and literature.

To carry out the Lethbridge experiment, it was necessary to secure the co-operation of the local station CJOC. The Lethbridge station was chosen because already during 1937 it had made its facilities available for six months for a daily "School of the Air" which had been heard by more than seventy rural and a few urban schools in the area as well as by schools across the American border in Montana. The success of these programmes gained the approval of the Deputy Minister, Dr. G. F. McNally, and paved the way for the implementation of the Committee's plans. Mr. G. Gaetz, the manager of station CJOC, readily agreed to give the station's facilities during the experimental stages of the work, with the proviso that as soon as the Department of Education decided to extend the scope of its programmes to cover the province as a whole, he would discontinue his local "School of the Air" in its favour.

In the light of the favourable reports it received from teachers regarding the first year's work at Lethbridge, the Committee on Radio Education decided to take a further step. It approached Mr. Gladstone Murray, General Manager of the Canadian Broadcasting Corporation, seeking his help in creating a network of Alberta stations. An encouraging reply was received.

First Alberta Network Series

At the Committee's meeting in Edmonton on October 15, 1938, it was decided to launch a series of school broadcasts running from November 14 to April 30 over an Alberta educational network which was to include CKUA Edmonton, CFCN Calgary, and CJOC Lethbridge. Three series were undertaken. The first was a weekly news broadcast for social studies classes, given by Mr. Watson Thompson, a lecturer on the staff of the Extension Department of the University of Alberta. The second was music for the intermediate schools, given jointly by Mr. G. Jones, F.R.A.M., and Mr. T. Jenkins, Ms.B., both of Calgary. The third series, music for the elementary school, was given by Miss Janet McIlvena, Supervisor of Music for the Lethbridge schools.

A survey made by Dr. Newland indicated that approximately 150 radios were in use in Alberta schools; but it was agreed that the type of receiver ought to be carefully considered and that some standard class-room receiver ought to be chosen by the Committee and recommended for use.

On Saturday, July 29, 1939, the CBC opened its new high-power transmitter (50,000 watts) at CBK Watrous, Saskatchewan, covering the greater part of the three Prairie provinces. (Incidentally, the "K" in CBK is said, by the Saskatoon *Star-Phoenix*, to commemorate Henry Kelsey, the eighteenth-century explorer who first saw near Watrous the great herds of buffalo ranging over the prairies.) The opening ceremony was marked by speeches from the Honourable C. D. Howe, Mr. Leonard W. Brockington (Chairman of the CBC Governors), Mr. F. W. Ogilvie, the Director-General of the BBC, Mr. William S. Paley, President of CBS, and the premiers of the three Prairie provinces. Perhaps the speeches on this occasion, which emphasized the vast coverage of the new transmitter over most of the three Prairie provinces and well into the United States, gave Dr. Newland the idea of widening the scope of school broadcasting through the use of CBK. He believed that Alberta should take the lead in persuading the departments of education of Saskatchewan and Manitoba to undertake a joint regional programme of school broadcasts. Accordingly Dr. Newland made approaches to these departments, but had to report that neither was yet ready to take part in such a co-operative venture. Alberta therefore decided to continue its own programmes by itself for another year.

By this time, school broadcasting was becoming well established among the teachers of the province. Several hundred favourable reports were received from both city and rural schools, emphasizing in particular that the broadcasts were "a great boon to pupils in rural schools."

At the next meeting of the Committee, in August 1940, the question of paying an adequate fee to the broadcasters was brought up. The Committee decided that a standard rate of remuneration for each programme should be established of $10 per script. The subjects proposed for the school year 1940–41 were music for primary and intermediate grades, social studies for intermediate grades, and science for elementary and intermediate grades.

CBS "School of the Air"

At the same meeting, the Committee gave consideration to a proposal from the CBC offering two series of broadcasts from the Columbia School of the Air,[1] made available in Canada through the courtesy of CBS. The subjects of the two series offered were music (folk songs) and literature ("Tales from Far and Near"). It was decided to accept this offer and to schedule the two CBS series in addition to the four Alberta series previously approved, making a total of four half-hour and two fifteen-minute broadcasts each week. The Committee recommended that in no school should the time given to broadcasts exceed half an hour a day.

For this season, for the first time, an annotated schedule of the programmes was prepared in mimeographed form and distributed by the Department of Education to the teachers. It paved the way later for a printed booklet giving more detailed advance information on each programme.

On December 11, 1940, the efforts of Dr. Newland to interest the

[1]At various times from 1936 to 1946 CBS changed the title of its "School of the Air." It was known as "Columbia School of the Air," "School of the Air of the Americas," "American School of the Air," or "CBS School of the Air."

other Prairie provinces in school broadcasting bore fruit. On the invitation of the Honourable Ivan Schultz, Minister of Education for Manitoba, officials from the departments of education of the three Prairie provinces and British Columbia met at Saskatoon to discuss co-operation in school broadcasting. Manitoba was represented by H. R. Low, Superintendent of Education, and H. B. Hunter, Saskatchewan by A. B. Ross, Alberta by Dr. Newland, and British Columbia by Kenneth Caple, who had recently been appointed jointly by the CBC and the Department of Education to be Director of School Broadcasting in British Columbia. Mr. Andrew Cowan of CBC Winnipeg was also present for consultation.

Co-operation of the Four Western Provinces

At the meeting the representative of each province described what had been done in school broadcasting in his province. It was then agreed that the main objectives of school broadcasts were as set forth in the annotated schedule of Alberta school broadcasts: (1) that every child should learn to listen to broadcasts and to appreciate and evaluate them; (2) that though the radio can never displace the class-room teacher or class-room activities, it can guide, stimulate, intensify, and supplement class-room effort, especially in underprivileged communities; (3) that the radio should encourage interest in concerns of the communities and of the world outside the class-room, and foster in pupils that sense of civic and social responsibility on which rests the future of democracy.

In the discussion that followed, the conference endorsed Alberta's view that "school broadcasts prepared for local needs in the province and approved generally by the schools using them, should not be supplanted by others from outside the province." This proviso guarded against the possibility that either Prairie regional programmes or programmes offered on a national basis by or through the CBC should be used to displace the local provincial product. Having accepted this safeguard, the four departments of education agreed to experiment with a limited number of broadcasts supplementing provincial programmes but prepared co-operatively. Two fields were recommended as common ground for all four provinces, primary music and dramatized literature for grades 7 to 9. The CBC was asked to assist in preparing and presenting the broadcasts, and also in securing through its facilities satisfactory coverage over the whole area.

It was agreed that one of the co-operative series, primary music, should originate from the Winnipeg studios of the CBC and the other, dramatized literature, from its Vancouver studios. Each programme would be of thirty minutes' duration and the scripts were to be approved by a committee representing the four provinces. The expenses of script, acting, and other talent (estimated at $2400) would be shared equally by the four departments of education, while the CBC would provide production facilities and network lines.

At a further meeting of the Committee in the spring of 1941, the final details of the outline plan were worked out, and it was put into operation in the fall. To sum up, the Alberta school broadcasts during 1941–42 consisted of four types of programmes. Two series of Western school broadcasts, produced in co-operation by Manitoba, Saskatchewan, Alberta, and British Columbia, were presented over Alberta stations CFRN Edmonton, CFCN Calgary, and CFGP Grande Prairie, in addition to CBK Watrous; the two series were "Highways to Adventure" (dramatization of books for intermediate pupils) and music, under the title "Mother Goose and her Music" (for primary grades) and "Alice in Melodyland" (for junior grades). Two series of programmes of the CBS School of the Air were similarly carried over stations of the CBC midwestern network, as was a special series on child guidance, entitled "The Child in Wartime." Under this arrangement CKUA Edmonton limited itself to carrying two series of school broadcasts specially intended for reception in Alberta only. These consisted of broadcasts on oral French and current events, intended for Correspondence School students.

Correspondence School Broadcasts

During the four years following 1941, the Correspondence School Branch of the Alberta Department of Education experimented with a number of series broadcast in out-of-school hours over CKUA. These were at first directed to their first-year French students and grade 9 social studies students. Later all high school correspondence students taking French were required to listen to the assignments given over the air and complete them as part of their regular work. Two other series "These Make History," a biographical series for social studies students, and "Choose Your World," a series of vocational programmes, were developed during the years 1942–44.

In 1945, Correspondence Branch staff writers turned their attention to programmes in English, music, current events, and science, all broadcast in the evenings. They were not only written but also directed and performed by the staff using the facilities at CKUA. In 1946, two of the Correspondence School series, "Through the Magic Door" and

"Speech Training," were presented in the afternoon during school time.

When the School Broadcasts Branch of the Department of Education was set up in 1946 it was housed in Correspondence Branch offices and staffed by two Correspondence School teachers, Doris Berry as co-ordinator and Helen Macmillan as script editor. In 1948 the School Broadcasts Branch was reorganized and was fully established as part of the Curriculum Branch of the Department of Education.

By then the number of Alberta schools on the Department's radio mailing list had risen to 750, and the Committee declared itself satisfied "that the school broadcasts perform a valuable service, especially in small schools in the outlying areas of the province." Its opinion was reinforced by the reports sent in to the Department by teachers who used the 1941–42 programme series. The most popular broadcast was the series of junior music, the reason given being that it stimulated pupil activities. "Any broadcast outside of stories," remarked one of these teachers, "must provide activities for the children or the children lost interest."

The chairman of the Radio Committee, Dr. Newland, was at this time still sceptical about the value of school broadcasting in general. He knew that the programmes were popular with the pupils themselves but preferred to take the verdict of teachers on the matter. "Teachers," he said, "judge the contribution made by school broadcasts not from their entertainment value, but rather from their direct use in class-room activities." He admitted, however, that broadcasts of stories or social studies dramatizations might also have a stimulating effect.

Dr. Newland was often heard to remark that he himself did not listen to radio school broadcasts. He preferred other educational media. "A school broadcast," he told the Alberta Radio Committee in 1943, "cannot be justified unless it brings to the activities of the class-room advantages which cannot be found in any other way for the same amount of time and expense . . . on this view it is a question of whether school radio has as much value as other media for this purpose." Dr. Newland referred specifically to the use of class-room and other library facilities.

National School Broadcasts Questioned

It is not surprising, therefore, that Dr. Newland was doubtful about the possibility of developing school broadcasts on a regional and national basis as distinct from a provincial basis. During the winter of

1941–42 he met Mr. Kenneth Caple, Director of School Broadcasts for British Columbia, and Mr. Andrew Cowan of CBC Winnipeg, to discuss the Western school broadcasts produced in co-operation by Manitoba, Saskatchewan, Alberta, and British Columbia. Dr. Newland said he was not satisfied with the results of the co-operative arrangement. He doubted whether either a national or an interprovincial series of school broadcasts could be presented without "generalizing" the broadcasts to the point where they ceased to have much local interest or value. He admitted, however, that the objection did not apply with equal force to all subjects suitable for broadcasts, and in particular to music and language broadcasts. He also complained that the interprovincial series were too expensive, mainly because they required the employment, on the CBC network, of professional musicians, actors, and script-writers. These, Dr. Newland held, might add to the entertainment value of the broadcasts, but local broadcasts given by teachers themselves had more educational value.

In spite of these objections, the Radio Committee agreed to continue for another year with one series presented co-operatively by the four Western provinces—namely that entitled "Canadian National Life." It also decided that all Alberta school broadcasts should henceforth come under the direction of station CKUA and originate from that station, if possible, provided the use of the telephone lines could be procured. The Committee agreed that the afternoon period was the best time for school broadcasts, and that all schools in the province ought to receive the annotated schedule of the programmes.

In 1943 the position of CKUA again became a matter of concern. Owing to war restrictions, there was great difficulty in securing adequate accommodation for school broadcasts on the telephone lines of the province. An appeal was made to Dr. James Thomson, the General Manager of the CBC, requesting that CKUA be equipped to serve as a regional outlet of the CBC. After further discussion, however, this idea was found to be not feasible, and in 1945 the station passed from the control of the University into that of the Alberta government telephone system.

Dominion Board of School Broadcasts Suggested

In 1943 the Committee had before it a plan presented by the writer, then Educational Adviser of the CBC, for a series of nation-wide school broadcasts planned and produced by the CBC. The Committee's chairman was decidedly sceptical about this proposal. "In whose hands," asked Dr. Newland, "should be placed the control of any programme of school broadcasts designed to reach a million Canadian school children?"

At the same time Dr. Newland admitted that there was a real need for a series of broadcasts based on recognition of the principle that the post-war reconstruction of democracy would be possible only if boys and girls in the intermediate and high school grades could be trained to understand the social forces then at work, and to see the possibilities for reconstruction through the freer use of science, technology, and the natural resources of the country and people of Canada. Dr. Newland made the practical suggestion that "a Dominion Board of School Broadcasting" should be set up to consist of representatives of the provincial departments of education. This Board should have the responsibility and authority for preparing and arranging for national school broadcasts. The same principle, he considered, might be extended to interprovincial or regional school broadcasts.

On July 6, 1943, Dr. James Thomson, having considered the suggestions made by Dr. Newland and his Alberta Radio Committee, replied to them as follows:

> Education is organized on a provincial basis, whereas the Canadian Broadcasting Corporation recognizes fully that the direction of education is in the hands of provincial departments and we have no right to intrude upon what is strictly a curriculum responsibility for the authorities concerned. On the other hand, radio broadcasting is a technical matter, not only from the point of view of actual physical apparatus required, but also from the point of view of presentation. There is growing up a body of experimental knowledge in connection with radio broadcasting, and in the last resort the Canadian Broadcasting Corporation must be responsible for every programme that goes over the air.
>
> Some of these problems may appear somewhat difficult and intractable when looked at from the theoretical point of view, but as a matter of fact, their solution is not difficult if we begin from the point of view of co-operation. Where there is goodwill and a desire to work together, there is no reason why the Canadian Broadcasting Corporation and the various provincial departments should not come to a very happy and harmonious working arrangement for doing this important public service. So far this has worked out very well, and I see no reason why it should not continue for the future.
>
> I am interested in your suggestion about a National Co-operative Committee. As a matter of fact, we have this in mind, and I expect we shall move into some arrangements in the near future.

Dr. Thomson's reply foreshadowed the action which took place at the annual convention of the Canada and Newfoundland Education

Association in Quebec in September 1943 where the delegates agreed to a plan for the National Advisory Council on School Broadcasting.

Western Regional Committee Set Up

At a meeting convened in Banff, Alberta, on August 3, 1943, Alberta agreed to take part in a Western Regional Committee for School Broadcasting, consisting of representatives of the four Western departments of education and of the CBC. At this meeting it was also agreed that there should be not more than one national series and one regional series of programmes per week.

Each department of education represented on the Committee was asked to set up a "panel" of experts, which was to study the curricula and techniques of instruction in all four provinces for the purpose of discovering common elements that might be dealt with in school broadcasts. Alberta proceeded to nominate experts in social studies, music, literature and dramatics, health, and science. However, it proved in practice difficult to arrange for joint meetings of experts from all four provinces, and this piece of interprovincial research was later abandoned in favour of meetings of the directors of curricula of the four provinces with the members of the Western Regional Committee for School Broadcasting.

In 1943–44 three series of programmes from the CBS School of the Air, as carried by stations CBK, CFRN, and CJCA, were included in the Alberta school broadcasts. These were, besides "Tales from Far and Near" (dramatizations of children's books), a geography series "New Horizons" and a science series "Science at Work." However, these programmes were criticized on the ground of their lack of correlation with the Alberta curriculum. Accordingly they were dropped from the regular schedule of Alberta school broadcasts for 1944–45, although they continued to be scheduled for out-of-school hours listening until 1947–48.

For 1944 and the following years the Alberta schedule was reorganized into two distinct groups of offerings. One group consisted of programmes heard over a network of stations arranged by the CBC, and produced under the supervision of the CBC regional office in Winnipeg. They lasted from 10:30 to 11:00 A.M. each school day, and included the national school broadcasts and the series presented by the four Western departments co-operatively. The second group consisted of other series planned by the Department of Education but presented, generally in afternoon periods, over local stations (including CKUA) rather than over CBC facilities.

Outstanding Alberta Series

Foremost among these was the elementary music series "Sing and Play" given by Miss Janet McIlvena, L.R.S.M., of Lethbridge. Her programmes originated from the studio of the Lethbridge station, and continued until her death in 1958. According to the Department of Education, "probably no single individual had a more profound influence on the early musical training of young school children than Janet McIlvena." Her series and another series on speech training were heard in an early afternoon period. The broadcasts to correspondence school students (in oral French, and vocational guidance) continued to be produced from CKUA in late afternoon or early evening periods. Another good music broadcast was developed by Miss Hazel Robinson, script editor from 1950 to 1954, for elementary grades, entitled "Music Makers." Miss Robinson also planned the first series of school broadcasts on creative writing, which proved very successful. Another outstanding contributor to Alberta school broadcasts was Mrs. Elsie Park Gowan, who gave up her teaching career for a time to become a popular and successful free-lance script-writer for the CBC nationally.

In 1945 Dr. H. C. Newland, who had played a dynamic role in the initial period of Alberta school broadcasts, left the Alberta Department of Education, and was succeeded by Mr. Morrison L. Watts, the Department's Director of Curriculum. He also replaced Dr. Newland as Alberta's representative on the National Advisory Council on School Broadcasting. Mr. Watts took no less keen an interest in the development of school broadcasting, and contributed much to its growth in the next ten years.

Progress in Alberta

By the winter of 1945–46 the Alberta school broadcasting schedule was fully developed, on a basis which provided a greater amount of programming than in any other province. The Department was providing from thirty to forty-five minutes five days a week for elementary and intermediate schools, as well as thirty minutes in the late afternoon and from fifteen to thirty minutes in the evening from one to five days a week for high schools. The high school programmes, of course, were planned for home listening by students. The details of the schedule are given in Table I.

Most of the provincial series were produced in the studios of CKUA, but the distribution of the programmes over the greater part of the province not covered by CKUA remained "spotty," being dependent on the varying co-operation of private stations in Calgary, Lethbridge, Medicine Hat, Grande Prairie, and other centres.

TABLE I

BROADCASTING SCHEDULE 1945–46

Fall term 1945	Elementary school 10:30–11:00	Intermediate school 3:00–3:30	High school 5:30–6:00	High school 8:15–8:30
Monday	—	music	French 1	literature
Tuesday	science	—	French 2	music
Wednesday	—	speech training	—	social studies
Thursday	music	—	French 3	vocational guidance
Friday	national	—	—	science

Spring term 1946	Elementary school 2:00–2:30	Intermediate school 5:30–5:45	High school 5:45–6:00	High school 8:30–9:00
Monday	music	French 1	—	"Tales of Adventure" (CBS)
Tuesday	science	"Win your World"	—	—
Wednesday	language	"Today's Horizons"	French 2	"March of Science" (CBS)
Thursday	music	science	—	—
Friday	national	French 3	—	music (CBS)

In 1948 the Curriculum Branch of the Department of Education assumed the responsibility for administration of all school broadcasts, working in co-operation with the Correspondence School Branch (under Mr. G. F. Bruce) which undertook to supply the scripts. Henceforth most of the provincial school scripts were written by free-lance writers, some of whom were also teachers in the Correspondence School. Later Miss Doris Berry, of the Correspondence Branch, became the first Supervisor of School Broadcasts. She organized the evaluation of school broadcasts by teachers on a systematic basis. Their reports indicated that about 700 schools possessed radio receivers, and that about 500 schools were making regular use of the programmes. In 1955 Miss Berry left Edmonton and went to Toronto to join the staff of the CBC School Broadcasts Department. She was succeeded as Supervisor by Mr. Richard A. Morton.

The 1946 report of the Chairman of the Alberta School Radio Committee significantly summed up the progress which had been made. Ninety-two per cent of all teachers using school broadcasts, he stated, said that the broadcasts helped in the attainment of educational objectives. "These teachers are almost unanimous in their opinion that the broadcasts stimulate the pupils' interest to a marked degree. They comment also on the training the broadcasts provide in habits of listening and desire on the part of the pupils to seek further information. One teacher remarks on the living quality which cannot be provided in books; another that the broadcasts are the equivalent of a visiting teacher who is an expert in his particular subject. Some teachers say that the broadcasts provide the only music experiences in their rural schools." Evidently, school broadcasting had by this time won an established place and proved its worth as a teaching aid in the Alberta school system.

SASKATCHEWAN

Saskatchewan, like other parts of Canada, shared in the general enthusiasm for educational radio in the early 1930's. Here, this enthusiasm took practical shape in an experiment with the use of out-of-school radio broadcasts to supplement in-school teaching. In October 1931 the Government Correspondence School, under the direction of Mr. A. B. Ross, began broadcasting a daily series of lessons covering the high school courses in English, history, science, Latin, and German. In subsequent years geography was added to the list of subjects. Each week the five half-hour programmes were all directed at one grade level of students, for example, the first week to grade 9, the second to grade 10, and so on alternately. The lessons, given in lecture form, were prepared by the instructors of the Correspondence School to supplement the work outlined in the course. Students' difficulties and questions were also discussed during the periods.

The broadcasts were, by arrangement with the Government Telephone Department, carried as a free public service from 6:00 to 6:30 P.M. over a network of four Saskatchewan radio stations, CKRM Regina, CFQC Saskatoon, CJGX Yorkton, and CKBI Prince Albert. They met with a favourable reception from parents, teachers, and students. One teacher, for example, reported that there were four listeners in her home—two adults over fifty and two children of grade 8 and grade 9 level respectively.

In 1932 the annual departmental report to the Minister stated that the broadcasts "provide an additional medium whereby students, who are struggling to get an education, receive direct assistance in their studies. They have given many an older person an opportunity to brush up studies. They have afforded means of enjoyment as well as profit to many who are shut off from the ordinary lines of active study.

Without a doubt this programme of radio instruction is achieving its purpose in supplementing the correspondence lessons."

Unfortunately, the Correspondence School experiment did not continue beyond 1935–36. Thereafter, crop failures and resultant curtailment of activities forced the Department of Education to drop the broadcasts. However, in 1938 the legislature amended the School Grants Act and the Secondary Education Act to give to school districts help in purchasing radio receivers, film projectors and the like, by a grant of 25 per cent (later increased to 40 per cent) of the cost. The grants were particularly helpful to the smaller rural areas of the province.

Impetus from the CBC

The next impetus came from the CBC. All the Prairie provinces were naturally influenced by the successful establishment of school broadcasting in British Columbia between 1936 and 1941 as the result of direct co-operation between the Department of Education of that province and the CBC. Upon its own initiative, the CBC, during the winter of 1940–41, decided to carry at its own expense two of the five daily series of the British Columbia school broadcasts (originating in Vancouver) each week over its Prairie transmitter. These programmes were listened to with appreciation in many Prairie schools and aroused so much interest among the departments of education that in December 1940 the Honourable Ivan Schultz, Minister of Education for Manitoba, suggested that a joint meeting of the four Western provinces be held to consider further co-operation.

Western Provinces Decide to Co-operate

The meeting took place at Saskatoon on Wednesday, December 11, with A. B. Ross representing Saskatchewan. The meeting agreed on the objectives of school broadcasts (see p. 55) and on the importance for local needs of school broadcasts prepared in the province as compared with others produced outside the province. The meeting recommended the joint preparation of a few broadcasts to supplement the provincial programmes, the expense of production being borne equally by the four departments, and the scripts being supplied by a committee representing the four provinces.

In order to carry out in Saskatchewan the recommendations of the meeting in Saskatoon, it became necessary for the Saskatchewan Department of Education to establish a branch which could organize school broadcasts in the province. Thus the Audio-Visual Instruction Branch of the Department was set up in the fall of 1941 with the following aims: to cover the distribution of audio-visual aids, the preparation, development, and sponsoring of educational broadcasts, the supervision of the rural circuits of the National Film Board, the administration of use of class-room national school broadcasts and audio-visual aids to school boards and other school organizations, and the issue of teaching outlines to help teachers make effective use of these aids.

Appointment of Mr. Morley Toombs

Chosen to head the new branch was Morley P. Toombs, Superintendent of Schools, who had been working on curriculum developments in the province and had taught a class at the Provincial Summer School in 1941. Mr. Toombs was enthusiastic about the prospects for a national CBC series of school broadcasts, which was then under discussion.

Early in the fall of 1941 Mr. Toombs was invited by the Deputy Minister of Education, Mr. J. M. MacKechnie, to organize and to develop school broadcasts in co-operation with the other Western provinces. As a result, the first school broadcasts were heard in Saskatchewan over CBC and privately owned stations from October 1, 1941, to April 30, 1942. There were two series—on Tuesdays a language-arts series entitled "Highways to Adventure" for grades 5 to 10, and on Fridays a junior music series for grades 1 to 4. Radio guides were prepared by the Branch and distributed to all schools known to have radios (approximately thirty-three) and to other schools upon request. The Department of Education's policy of offering grants of 40 per cent of the cost of projectors and radios to schools desiring to purchase such equipment (up to $25 per school) gave a powerful impetus to the purchase of school radio sets, and the number of listening schools rapidly increased.

Saskatchewan Supports National Series

Towards the close of this first year's work, the writer of this book visited Regina and attended a departmental meeting held in the Legislative Building where he presented proposals for a forthcoming series of national school broadcasts to be produced at the expense of the CBC, with each department of education contributing a script of its own choosing, the title of the series to be "Heroes of Canada" (17 broadcasts). This proposal was received with enthusiasm by Saskatchewan, which undertook to contribute one programme to the series entitled "Angus McKay—Protector of Wheat." Later the programme was written by a local script-writer, Miss Rowena Hawkings, and produced on December 4, 1942, in CBC's Winnipeg studios by a leading drama producer, Mr. Esse W. Ljungh. During the winter of 1942–43 the Saskatchewan schools again received two programmes per week, junior music, a joint presentation of the four Western departments of education, and "Heroes of Canada," a national series in which all departments of education and the Canadian Teachers' Federation participated. An estimated 500 Saskatchewan schools listened to these broadcasts.

Interest continued to grow and on April 13, 1943, a special school radio conference was held in the Legislative Building in Regina under the joint auspices of the CBC and the Department of Education. To this meeting were invited (besides CBC representatives and Department of Education officials) representatives of the Saskatchewan Teachers' Federation and the managers of various privately owned radio stations in the province. The CBC representatives on this occasion were Mr. George Young, Mr. Dan Cameron, and the author.

As a result of an appeal to the CBC and the Saskatchewan private stations, it was agreed that, during 1943–44, daily school broadcasts should be made available to Saskatchewan schools from 2:00 to 2:30 P.M. MST, and that these programmes would be carried by a complete network of stations in the province. This schedule of programmes was to include not only the CBC national series and a joint presentation of the four Western provinces, but also two series from Columbia's School of the Air and a series to be prepared by the Saskatchewan Department of Education in co-operation with the Saskatchewan Department of Public Health. The latter series, entitled "Health Highways," was the first Saskatchewan series to be written within the province. To undertake the work Mr. Toombs employed a young script-writer from station CKCK named James W. Kent. The originality of his scripts soon attracted attention and general appreciation. The number of listening Saskatchewan schools was now estimated to be 500, and the Department reported "that the use of school broadcasts is becoming a definite part of the teaching techniques of many Saskatchewan teachers."

In the summer of 1943 a meeting of the Western Regional Committee for School Broadcasting was held in August at Banff, Alberta.

Here plans for continued and strengthened co-operation were worked out between the departmental and CBC representatives.

During the following winter from October 1943 to April 1944, the schedule of Saskatchewan School Broadcasts was as follows:

Monday: "Science at Work" (CBS School of the Air)
Tuesday: junior music (four Western provinces joint production)
Wednesday: "Health Highways" (Saskatchewan's own series)
Thursday: "Tales from Far and Near" (CBS School of the Air)
Friday: CBC national series, including "Serving Canada," "The Adventures of Canadian Painting," and "Canadian Literature"

A teachers' manual giving full details of the programme was issued to Saskatchewan schools in October and January. Teachers were also provided with evaluation sheets to report their opinions of the presentation.

Reorganization in 1944

During 1944, considerable changes in the organization of school broadcasts were made in Saskatchewan. A change of government took place in June, as a result of which the former Liberal administration gave place to a CCF administration. A new Minister of Education was appointed, the Honourable W. S. Lloyd, a former teacher with strong sympathy for progressive developments in education. Mr. Henry Janzen became Director of Curricula and Dr. H. C. Newland (formerly of Alberta) Director of Research. Mr. Janzen's duties included responsibility for the audio-visual work, in whose development he took a keen and continuing interest. At this time Mr. Morley Toombs accepted an invitation from the National Film Board to become manager of its local film circuits and left the Department. Thereafter the Audio-Visual Branch was split into three sections, one for films, one for radio, and one for adult education. Mr. Leon Thordarson, formerly assistant to Mr. Toombs, continued to co-ordinate the three sections (under Mr. Janzen's supervision) during the winter, after which he returned to regular teaching work. In February 1945, Mr. E. F. Holliday was appointed Superintendent of Visual Education (films), while about the same time Mr. J. W. Kent, former script-writer from CKCK, became Supervisor of School Broadcasts. The adult education section was headed by Mr. Watson Thompson.

From October 1944 to April 1945 Saskatchewan schools continued to enjoy the daily half-hour school broadcast at 2:00 P.M. MST from Mondays to Fridays. They were carried over CKRM Regina, CHAB

Moose Jaw, CFQC Saskatoon, CKBI Prince Albert, and CJGX York-ton. By this time the CBS School of the Air programmes were no longer being heard in Western Canada, and were replaced by programmes produced either provincially or regionally, the schedule being as follows:

Monday: "Adventures in Speech" for primary grades (originating from Winnipeg after Christmas)
Tuesday: "Science for Today" for grades 5–9 (a co-operative series of Saskatchewan, Alberta, and Manitoba)
Wednesday: "Adventures in Modern Living" for grades 3–10 (Saskatchewan series produced in co-operation with the Department of Health)
Thursday: intermediate and senior music for grades 5–9 (a co-operative presentation of the four Western provinces)

The total number of radio-equipped schools was now estimated to be 600 and the Department gave consideration to the desirability of equipping all Saskatchewan schools.

Music Broadcasts

In 1945 the Department began providing a new type of school music broadcast, which reflected the distinctive philosophy of its recently appointed Music Supervisor, Mr. Rj Staples. He laid great stress upon helping children to respond to music through free physical movement and through making and playing their own instruments. This philosophy has affinities with the well-known Carl Orff system of musical education in Germany, and with the ideas of Miss Ann Driver in England. Mr. Staples was specially concerned with familiarizing teachers throughout the province with his new approach to music. Accordingly the music broadcasts served not only to stimulate the children in the class-room, but also as model lessons showing the teacher how to teach music.

Mr. Staples developed what he called "a pattern approach to music reading" based upon an adaptation of procedures from the learning and teaching of spoken language to the field of music. Two principles are basic: first, that the child must have a wide experience in understanding music, and in expressing himself musically (through bodily movement, voice, and instruments) before he can be expected to read intelligently the symbols of music (notation), and secondly, that a limited rhythmic vocabulary (only six different two-measure patterns) must be used for the introduction of music reading. Accordingly, the primary purposes of the school music broadcasts have been to present a simple graduated approach to music reading for voices and class-

The National Advisory Council on School Broadcasting in session in 1959. This picture shows part of the council, including (*at the back*) the chairman, Dr. W. H. Swift (*right*), and the secretary, R. S. Lambert (*left*), and the following members, *clockwise from left*: Dr. Lloyd Shaw (P.E.I.), Mr. C. H. Aikman (Que.), Miss Gertrude McCance (Man.), Mr. Gerald Nason (C.T.F.), Mr. J. W. Grimmon (Ont.), and Mr. Ralph Kane (N.S.).

Children assisting in a British Columbia junior music school broadcast, "Alice in Melodyland," in 1941.

A demonstration of the School of the Air of the Americas held at Harbord Collegiate
Institute, Toronto, in May 1941. One-half of the platform is occupied by a
demonstration broadcast, the other by a class of children with their teacher
receiving the broadcast. Seated at right is Mr. Sterling Fisher, Director of the
School of the Air.

Planning the first music appreciation series in the Ontario school broadcasts, 1943.
Standing, left to right: George Dixon, CBC producer; Paul Scherman, violinist
and orchestra leader. *Seated, left to right*: Dr. Leslie Bell, Ontario Department of
Education; Dr. Roy Fenwick, Ontario Music Supervisor; R. S. Lambert, CBC
Supervisor of School Broadcasts.

Miss Irene McQuillan and a group of children taking part
in a Junior Music broadcast to Maritime schools.

Planning a national school broadcast in 1951, showing musician Louis Applebaum
at the piano, with (*left to right*) script-writer Earl Grey, CBC Supervisor R. S.
Lambert, Assistant Supervisor T. V. (Vick) Dobson, and producer Lola Thompson.

Mrs. Dorothy Adair, Grade I teacher at Givins School, Toronto, listens with her class to "Kindergarten of the Air" in 1950.

A class practising art expression in the CBC's Winnipeg studios during a broadcast of "It's Fun to Draw," under the supervision of Miss Elizabeth McLeish and Miss Gertrude McCance.

room instruments, and to combine the three phases of musical activity, bodily movement, singing, and the playing of class-room instruments.

The broadcasts provide a flexible pattern of activities for various grade levels. For example, the primary grades sing songs, do singing games, and play rhythm-playing-pattern on rhythm instruments. Intermediate and senior grades participate in the singing, play the melody and harmony instruments, and enjoy most of the dances. From time to time, other types of music broadcasts were featured in the Saskatchewan school broadcasts, including musical selections sung by adults, choral singing by school choirs, and the use of specially prepared recordings. Three albums of these records were published for the Department by Canadian Music Sales, Inc., and were regularly used on the broadcasts.

Appointment of Miss Gertrude Murray

In 1950 Mr. James Kent left the Department of Education to take up a position as drama producer with the Canadian Broadcasting Corporation. He was succeeded by Miss Gertrude Murray, who was responsible for the development of many of the newer features of the Department's schedule. Miss Murray worked in close association with the School Music Branch of the Department, and from 1951 to 1953 many of the programmes were prepared in a small "studio" in Mr. Rj Staples' offices. Magnacord equipment was purchased, which made it possible to prepare good quality tapes and send them to Winnipeg for presentation on the air, with the approval of the CBC. Other programmes beside the music series were prepared in this "studio," including "Creative Dramatics" conducted by Mrs. M. E. Burgess and "Rhythmic Playtime," planned and given by Miss Murray herself. The purpose of the programmes was to assist teachers to develop in young children an enjoyment of creative rhythmic activities. The three years during which this "studio" was used were important to the School Broadcasts Branch, giving it practical experience in the writing and production of programmes, an experience which proved of great value when in 1954 the CBC made its newly opened Regina studios available for the production of Saskatchewan school broadcasts.

Creative Writing

Another outstanding series of school broadcasts introduced by Miss Murray in 1954 was one entitled "Let's Write a Story," intended for students in grades 4, 5, and 6. The aim was to tell stories in such a way as to stimulate the children to express their own ideas in writing.

According to the Department's teachers' guide *Young Saskatchewan Listens* (1958–59), children "by learning to express their ideas in their own way will develop their imagination, improve their skill in handling their language, and become aware of the importance of writing in their life. The expression of their own thoughts provides personal satisfaction and helps a child to gain the respect of his classmates." The broadcasts not only tell stories, but discuss the use of language (choice of words, phrases, and sentences), the choice of plots and development of ideas. Listening classes were encouraged to send in stories and poems written by their members, which were read over the air on the final broadcast of the series. Miss Murray has reported that "the children themselves write freely and happily, knowing that we are more concerned with ideas than with mechanics. It has become an enjoyable experience to write and listen to 'our own stories'. Teachers tell us there is quite a carry-over into other subjects. The development of 'a flow of words' is noticeable in other forms of written work. Weaknesses revealed in their story-writing often forms the basis for many purposeful lessons in written language for other occasions." Excerpts from the work turned out by writing classes have been included by The Macmillan Company of Canada in sections on creative writing in language texts they have published.

Special Jubilee Presentation

In 1955, the Jubilee Year of the birth of the Province of Saskatchewan, school broadcasting played a considerable part in the celebrations. Three of the provincial series of programmes were planned to emphasize provincial history and school participation. "Saskatchewan Cavalcade," for example, presented twelve dramatized broadcasts covering early life on the plains, the fur trade, exploration, and the development of settlements.

Miss Murray and Mr. Rj Staples, who had collaborated before in the music field, prepared a special booklet containing commemorative jubilee songs composed by Rj Staples, Neil Harris, and others, which were taught on the broadcasts and formed the music section of a special tribute presented by radio to all schools on May 26, 1955. On this day in individual class-rooms, schools, and groups of schools, pupils gathered around class-room radios, in auditoriums, and school yards, where they followed the pageant presented in the previous broadcast and supplemented it with their own community pageants. The total audience was estimated at 100,000 and the songs continued afterwards to be used as "Saskatchewan songs" in the schools.

MANITOBA

As far back as 1925, the Manitoba Teachers' Federation began experimenting with education broadcasts. The programmes were general in scope, of half an hour's duration each, and were given after school hours. Representative teachers were invited to prepare talks on such subjects as "The Causes of the American Revolution," Drinkwater's play *Oliver Cromwell*, and so forth. According to one of the participants, Miss Aileen Garland of the Manitoba Teachers' College, "The broadcasts were useful as a review for students of subject matter dealt with at greater length in class, and especially useful for teachers who had little time for additional reading, and few books."

The favourable reception of these programmes led to their continuance for several winters. In 1928 the Department of Education showed its interest by sharing the responsibility with the Teachers' Federation. Gradually, the broadcasts were slanted towards a closer and more direct relationship to the provincial course of studies. From 1931 the Department took over sole control of the "radio school" and, in the light of experience, decided to substitute two fifteen-minute lessons on different subjects for the original half-hour talk. Plans for putting this into effect and transferring the broadcasts to school periods were ready in the fall of 1935, but were postponed because of financial difficulties resulting from crop failures.

In 1937 the project was revived. During the fall term a series of broadcasts designed to help teachers with their work by providing supplementary material in music, geography, history, science, and English was given. While it was in progress, secondary school teachers throughout the province were invited to say whether they wished to have broadcasts during school hours and, if so, of what kind. The replies received were favourable to the introduction of school broadcasts, but divided over the question whether the programmes should be of a general inspirational and supplementary character, or should be related to specific detailed lessons.

Accordingly, early in 1938, the Department decided to make a fresh start with school broadcasts. The Canadian Broadcasting Corporation had recently come into being, and had established a network across the country. In Winnipeg it was making use of the facilities of CKY, then owned by the Manitoba Telephone System. Through the courtesy of station CKY, two half-hour periods from 3:00 to 3:30 P.M. CST on Tuesdays and Thursdays were allocated to the Department of Education. The Tuesday period was to be used for talks on general subjects, grouped in units of four. The Thursday period was divided into two fifteen-minute periods dealing with music, geography, history, science, and English. Committees of teachers were appointed to plan the content of the programmes. Outlines of the broadcasts, with suggestions for preparatory and "follow-up" work, were provided. Arrangements were made to have class-room teachers report on the results. About seventy schools, representing a class-room audience of about 5,500 pupils, reported to the Department that they listened to the broadcasts. In many cases the teachers rented or borrowed receivers for this purpose; in others, money was raised by voluntary efforts to buy receivers.

However, under the influence of the depression of the 1930's, the impetus behind this second start died away, and no further school broadcasts were given until after the outbreak of World War II.

"Music and Movement" Series

In 1940 the late Mr. Harry Low, who had recently come over from Scotland to take up the post of Superintendent of Education, tried to inject fresh enthusiasm and energy into Manitoba school broadcasting. After making a tour of inspection of the rural schools of the province, he came to the conclusion that life in many of these areas was dreary and sterile for the children in the primary grades. He believed that broadcasts such as those of Miss Ann Driver (which he had heard over the BBC in Britain), dealing with the rhythmic aspects of music, might help to bring vitality to the study of music at the primary level. Accordingly he asked Miss Ethel Kinley, the Supervisor of Music for Winnipeg Public Schools, to select two Winnipeg music teachers to write and broadcast programmes of a similar nature to those in Britain. The teachers chosen, Miss Elizabeth Harris and Miss Elizabeth Douglas, undertook (for an honorarium of $5 per show) to script and perform over CKY two series of programmes, one for grades 1 and 2, the other for grades 3 and 4. At first the broadcasts, entitled "Music and Movement," were of thirty minutes' duration; but, when experience showed that this was too long for the attention span of the students, they were shortened to fifteen minutes.

The producer of these programmes was Mr. Dan Cameron, a former teacher, who in 1942 joined the staff of the CBC. From the outset Mr. Cameron took a keen interest in the educational success of the series, in refreshing contrast to many professional broadcasters, who ranked the worth of school broadcasts in terms of their value as "show business."

During 1940 two new developments began to affect the situation in Manitoba. The first was the welcome given by the departments of education of many provinces to the CBC's offer to broadcast, over its network, some of the series offered by the CBS School of the Air. Mr. Low joined in welcoming this offer as affording a stimulus to school broadcasting in general. At the same time he stressed the equal importance of promoting school broadcasts on a provincial basis, to meet local curriculum needs.

Co-operation with other Western Provinces

The second development arose out of the CBC's decision to carry weekly, during the school year of 1940-41, two of the British Columbia series of school broadcasts over its Western network, and the interest aroused in the possibility of co-operation between Western departments of education. Manitoba's Minister of Education, the Honourable Ivan Schultz, took the initiative in proposing that representatives of the departments of education of the four Western provinces should meet together to discuss possibilities of co-operation in school broadcasting. The meeting was held on December 11, 1940, at Saskatoon (see p. 54), Manitoba being represented by H. R. Low and H. B. Hunter. There it was agreed that the four departments should co-operate in two series originating alternately from Winnipeg and Vancouver. The subject of the Winnipeg series would be primary music, and of the Vancouver series children's books (dramatized).

The two interprovincial series made a successful beginning during the winter of 1941–42. In Manitoba the broadcasts were heard from 3:00 to 3:30 P.M. CST on Tuesdays and Fridays each week. They were carried over CBK Watrous, CKX Brandon, and CKY Winnipeg. On the other three days of the week, three series of the CBS School of the Air (science, literature, and social studies) were carried by the same stations.

Manitoba thus had a complete schedule of school broadcasts available for use in its schools, but the only series originating from Winnipeg was that on primary music. In the Department of Education, responsibility for the school broadcasts was entrusted to the editor of the *Manitoba School Journal*, Mr. Max Warwyko, who worked closely with the CBC school producer, Mr. Dan Cameron. During 1942 the CBC showed what could be done in fields other than music by broadcasting an actual class-room situation from Fort Whyte school. The title of the programme, "How to Teach Latin," shows that it was intended primarily to provide a model lesson for the benefit of teachers.

In 1942–43 the work of school broadcasting was included in the duties performed by Mr. Edward Armstrong, who was also responsible for the use of educational films in Manitoba schools. As both these sides of his work were growing fast,[2] it soon became clear that one official could not satisfactorily cope with the two media. Mr. Armstrong decided to concentrate his attention on the film work, and accordingly the Department began to look for a Director of School Broadcasts.

Miss McCance Appointed Director

Miss Gertrude McCance, a graduate of the University of Manitoba, had specialized in speech training and had spent a year in London (England) studying under the well-known teacher Miss Marjorie Gulland. Returning to Canada, she taught in both public and private schools, mostly at the junior high school level. While on the staff of Havergal College, Toronto, she took part in the first speech course organized in Toronto by Mr. C. F. Cannon (later Chief Director of Education for Ontario). In the fall of 1943, Miss McCance returned to her native Winnipeg to become head of the Speech Department at Manitoba Teachers' College. At the wish of Mr. Schultz, she undertook to give part of her time to experiments with school broadcasting.

Miss McCance soon found herself increasingly attracted to the school radio work. She had a natural gift for broadcasting, and found in radio an ideal medium for extending the scope of her interests in speech training and choral verse speaking. Her first experiments were conducted in co-operation with the CBC and broadcast over station CKY. On January 10, 1944, she began a series of poetry readings for high schools which was followed by programmes given by the verse-speaking choir of the Normal School. A year later, in January 1945, Miss McCance started a speech training series for grades 4 to 6. Subsequently, the series was adjusted to the primary level, for grades 2 and 3 especially. It then became established as one of the most successful and widely heard series in Canada, under the title "Adventures in Speech."

In 1944 Miss McCance was appointed Director of School Broadcasts for Manitoba. She brought to the work notable qualities of drive, imagination, and organizing skill, all fused by a strong will that overcame obstacles, and sometimes even protocol. She enjoyed excellent

[2]In October 1942 the *Manitoba School Journal* stated that 100 rural and 20 city schools (9,000 pupils) were provided with radio receiving equipment, while the mimeographed programme bulletin was being distributed to more than 400 teachers.

relations with the successive ministers and deputy ministers under whom she served, and collaborated closely, if on occasion stormily, with CBC personnel. With Mr. Dan Cameron, her producer, she remained always on terms of mutual sympathy and understanding, a happy *entente* which continued for years after when Mr. Cameron had been promoted to Programme Director for the Prairie region.

The Manitoba Department of Education followed a broadminded policy of allowing its officials and teachers to appear at the microphone and participate personally in school broadcasts. On such occasions (except in the case of departmental personnel) they received suitable payment for their series. As a result of this policy, the field of specialized talent available for school broadcasts was used to the best advantage.

In 1944–45 and 1945–46 the first fruits of Miss McCance's appointment were seen in the schedule of Manitoba school broadcasts for these two years. The schedule, covering language, science, literature, and music was as follows:

Monday: "Adventures in Speech" for primary grades (study and appreciation of the English language)

Tuesday: "Science on the March" for grades 5–10 (two consecutive series on scientific thought and history of scientific achievement)

Wednesday: "Tales from Far and Near" for grades 7–10 (CBS School of the Air dramatizations of famous children's books)

Thursday: intermediate and senior music for grades 5–9 (co-operative series of four Western provinces comprising "Music Everywhere" [from Winnipeg] and "Listening Is fun" [from Vancouver])

All the broadcasts were presented from 11:30 A.M. to 12:00 noon CST. They were heard over the same network as before, except that in 1945 station CFAR Flin Flon was added by transcription.

Choral Music

Manitoba took very seriously its share of the Western provinces co-operative music series. The schools of Manitoba have a tradition of fine choral singing, due in part no doubt to the high proportion of new Canadians from Eastern and Central Europe in the population. Miss McCance therefore determined, with the assistance of Miss Ethel Kinley, to make school choirs a strong feature of the Manitoba broadcasts, especially at the intermediate grade level. Originally titled "Music Everywhere" the series was, after an experimental first year, renamed "Folk Music of Many Lands," its scope being extended to correlate folk music and social studies, and thereby widen the horizon of the school listening audience.

Miss Muriel James, a Winnipeg school music teacher, planned the series and, with the assistance of a colleague, Miss Margaret Thomson, wrote the scripts. She had the expert advice of a special committee of music teachers convened by Miss Ethel Kinley, Supervisor of Music for Winnipeg Schools (and after Miss Kinley's retirement, by her successors Miss Marjorie Horner and Mrs. Lola MacQuarrie). Songs and recordings were used on the programmes, and pupils in listening class-rooms were encouraged to participate in the broadcasts by singing, following simple rhythm patterns, and giving answers to questions.

Miss James conducted extensive research into the music and customs of the various national groups presented on the programmes. She made special trips to Wisconsin, Chicago, and Seattle to collect records suitable for her purpose. Choirs from Winnipeg and suburban schools were chosen by Miss Kinley to participate in the series and sing the illustrative folk songs. Mr. James Duncan, of Gordon Bell Schools, was chosen to narrate the series and comment on the folk songs. The first programme went on the air on January 6, 1947. In 1948 an innovation was made by the introduction on each broadcast of a "teaching song" to be taught by the narrator and practised by the listening audience. After 1948, with the departure of Miss James to the Pacific coast, her role as script-writer, soloist, and narrator was taken over by Mr. Duncan.

A further impetus to choral singing was given when Miss McCance persuaded the National Advisory Council on School Broadcasting to allocate the national programme on the Friday before Christmas each year to a performance of Christmas carols sung by a representative school choir. Each province, beginning with Manitoba, took its turn to provide this excellent demonstration of the singing achievements and standards of its school music.

"Adventures in Speech"

Naturally, the series "Adventures in Speech" under the impetus given it by Miss McCance, both through her personal performance and through her expert knowledge of the subject, achieved the widest fame of all Manitoba school broadcasts. It met a need, strongly felt throughout Canada, for finding some way to encourage correct English speech among school pupils at the primary level. The weekly broadcast turned out to be an ideal way. Miss McCance used poetry, speech rhymes, and stories to present amusing opportunities for developing and practising good speech habits. She preferred to address her class-room listeners directly, rather than bring a class or individual youngsters into the studio to share the lesson. However, she often made use of a local announcer to play the part of "stooge" in pronouncing words or repeating phrases, for the benefit of her audience. Occasionally she introduced a speech choir to demonstrate what she had been teaching.

Miss McCance tells many anecdotes of her "Adventures in Speech" series. Once, after a broadcast in which the sound "Moo" had been taught and practised, she heard from a little girl who wrote, "I live on a farm. I am in Grade Two. I listen to your programme every Monday. Last week our cow had a baby calf, and I want you to know she said 'Moo' just like you." On another occasion, after some programmes of poetry readings, Miss McCance heard from a prisoner serving in a penitentiary: "As I will be spending some time in this gaol, I would like to know what you think of my talents for writing poetry." Enclosed were a number of handwritten pieces which, on further examination, were found to have been copied from current magazines!

"Adventures in Speech" proved so popular in Manitoba that teachers in the neighbouring province of Saskatchewan began to ask for it. The series was first extended to Saskatchewan in 1945. From there, its fame spread steadily eastward. In 1950–51 it was first included in the Ontario and Quebec (Protestant) school broadcast schedules; and in 1952–53 in the Maritime schedule. The programme is still being heard in these eight provinces, covering four of the seven time-zones in Canada.

French by Radio

At its first meeting in 1944 the National Advisory Council on School Broadcasting set up a committee to investigate some of the problems involved in the teaching of French by radio. The committee recommended that radio French lessons should be divided into "instructional" and "enrichment" types of broadcasts, and that experiments which could be used as bases for further research should be conducted at local or provincial levels.

Following up this recommendation, Miss McCance formed a special committee consisting of members of the French Curriculum Revision Committee in Manitoba. This committee organized, through the CBC, a series of six fifteen-minute experimental broadcasts aimed at grades 8 and 9. The aims of the committee were to give students additional skill in oral work, to supplement the broadcasts with printed aids for use by the teacher, to use a variety of voices, to provide instruction in phonetics, and to include enrichment broadcasts of a simple lively and varied nature that would stimulate the children to a greater use of the language. The script-writer for the series was Mr. Alfred Glauser, of the Department of French of the University of Manitoba (now with the Department of French of the University of Wisconsin). He continued to write the French scripts till 1950, when the work was taken over by Madame Ragot, an experienced French teacher and a native Parisienne.

The experimental series showed that dramatization was the most effective method of presenting material in French. Narration of a formal type was not suitable. Phonetics, as taught on these programmes, was not successful. On the basis of this experience, French broadcasts for Manitoba were started regularly in 1946–47, under the title "Le Quart d'Heure français." The programmes were aimed primarily at grades 7 to 9, with some modifications enabling them to be useful also to students in grades 10 and 11.

Over the next six years, many different types of programme suitable for these levels were presented, and their relative effectiveness checked by teachers' evaluations. By 1952 the broadcasts had assumed their definite form, consisting of simple conversation, songs, drill in grammar and pronounciation, dramatizations of selections from texts, and occasional adaptations from French literature. To avoid stereotyping, however, each year some novelty in presentation was introduced. The basic method of preparing the script remained the same. Some idea of the careful spade work done by Miss Mary Reid, the senior assistant planner, can be seen from the following description of procedure:

First, a careful study is made of all texts from Grades 7 to 12 in the Manitoba curriculum. Any changes in curricula are noted. Next, texts with different ideas of presentation are also reviewed. Ideas for programmes are jotted down. Then a teacher and script-writer convene; twenty programmes are drafted; and text references noted to show the basis for the programmes. These suggestions are then submitted to Miss McCance who, in turn, submits them to a committee of teachers who give helpful criticism. Next, a teacher and script-writer choose ten programmes—the ones which seem most popular with the teachers' committee. Each program is discussed and the general form it is to take decided upon. The next step is the writing of the script. When completed, it is studied by the writer and teacher together to determine its suitability for the classroom. If necessary, the script is re-written. Finally the script is sent to Miss McCance, who decides if it meets school and radio requirements.

"It's Fun to Draw"

Probably the most popular and successful of all Manitoba school broadcasts has been the series on creative art, begun in October 1947 under the title "It's Fun to Draw" and continued ever since. The idea for this series came to Miss McCance through remembering her own dislike of the way she was taught "Art" when she was at school. In those days, a teacher would take a flower, stick it in a vase, and set it up before the pupils to draw. The result was a collection of conventional representations of an object in which the pupils took little or no interest or pride.

When Miss McCance was teaching at a northern Manitoba art institute for teachers, she met Miss Elizabeth McLeish and became enthusiastic over her new, more creative approach to art. Miss McLeish's method consisted of encouraging children to express themselves freely, without forcing them first to receive conventional instruction in drawing techniques. Following a visit to the schools of Brandon, Manitoba, where Miss McLeish was Supervisor of Art (later she became Director of Art at Manitoba Teachers' College, Winnipeg), Miss McCance resolved to experiment with the use of radio to stimulate creative self-expression among pupils in grades 4 to 9 in Manitoba public schools.

At Miss McCance's invitation, Miss McLeish worked out a series of programmes under the title "It's Fun to Draw." She explained to the teachers her plan:

Some of the broadcasts will employ drama, some will tell a simple story, others will present beautiful and interesting music. With this inspirational material, and some guidance in art principles, we hope to achieve some of the following aims:
  (1) To encourage children to be creative
  (2) To help children to express ideas fully
  (3) To foster an urge to create
  (4) To encourage an appreciation of children's own efforts and those of others
  (5) To aid children in getting the most out of their materials
  (6) To build up confidence in children's own efforts.

For the Department of Education, Miss McLeish prepared a special booklet for circulation to teachers advising them how to use the broadcasts of "It's Fun to Draw" in their class-rooms.

When Miss McCance heard that the title of the first radio programme in the series was to be "Creative Flowers," she felt moved to object, on account of her childhood memories. But she was reconciled when Mrs. Jean Edmunds, the script-writer, turned out an amusing and sparkling dramatization entitled "The Truth about the Dandelion —How its Hair Turned White." This, supported by music chosen by Mrs. Nancy Noonan, gave the listening audience an imaginative experience which stimulated at once a large variety of gaily coloured sketches. The early titles in the first series give a clear notion of the planners' inventiveness. They included "Hansel and Gretel," "Sing a Song of Sixpence," "Expressive Faces," "The Craft of Papier Mâché," and "At the Farm." "Sing a Song of Sixpence" evoked from one schoolboy a drawing of a table surmounted by a bottle of rye whiskey!

After each broadcast, hundreds of children's pictures of all kinds began to pour into Miss McCance's office. They were all carefully commented on by Miss McLeish. The children who had drawn and sent them in were encouraged to go on, with the help of constructive criticisms and helpful suggestions. The most interesting pictures each year were mounted and publicly exhibited. A special selection was reproduced on colour slides for showing to parents and teachers at Home and School meetings. Many parents were amazed at the talent displayed by their own children!

The fame of "It's Fun to Draw" soon extended far beyond the borders of Manitoba. The Saskatchewan Department of Education included the series in its schedule from 1948–49 onwards, and Alberta from 1950–51. The broadcasts were still continuing in all three Prairie provinces in 1960–61. In the course of time, articles appeared in Canadian magazines such as the Toronto *Saturday Night*, in the *British Empire Digest*, and elsewhere, describing "It's Fun to Draw" and praising the series as "the most important development in creative art on the Prairies" that had ever taken place.

Each year, the Manitoba teachers' guide to school broadcasts, *Young Manitoba Listens*, reproduced on its front cover in colour one of the best of the year's crop of pictures sent in by the pupils. Until 1945, the programme schedule of Manitoba school broadcasts had reached the teachers of the province only through being printed at regular intervals in the *Manitoba School Journal*. In 1946, however, a separate booklet was published for the first time, which became an annual publication (96 pages, illustrated), *Young Manitoba Listens*. The Department also published separate smaller booklets for "It's Fun to Draw" and scripts for "Le Quart d'Heure français."

From its shaky beginnings prior to 1945, school broadcasting in Manitoba had not only become firmly established, but by 1948 was also achieving considerable distinction nationally and internationally. In 1947 "Adventures in Speech" won the first of its two awards from the Institute for Education by Radio (Columbus, Ohio), and in subsequent years similar awards were given to "Let's Sing Together," "Le Quart d'Heure français," "Discoveries in Words," "Social Studies" (Japan), "Social Studies" (our American neighbours), and "Working

Together." Manitoba was the only province to receive one of the Canadian Radio Awards for its series of intermediate school music programmes.

There are several reasons for Manitoba's success, among them being the strong support given at the ministerial and curricular levels to the Director of School Broadcasts, vigorous leadership given by the Director, especially in the organization and execution of new ideas, support of teachers, through *ad hoc* working committees, close correlation of the broadcasts with curricular requirements, emphasis on student participation in the broadcasts, wherever possible, and happy relations between the Director and co-operating CBC personnel.

V. Ontario and Quebec

During the years preceding World War II and during its opening phases, the attitude of the Ontario Department of Education to school broadcasting was negative. On several occasions, Mr. Gladstone Murray, General Manager of the Canadian Broadcasting Corporation, made known the Corporation's willingness to provide network facilities for school broadcasting in any province where the Department of Education might wish to experiment with the presentation of school programmes. However, Dr. Duncan McArthur, the Minister for Education under the Liberal régime of Premier Mitchell Hepburn, indicated his view that expenditure of public money on school broadcasting would serve no worthwhile educational purpose.

In a large province like Ontario, the gap created by the negative policy of the Department of Education left the field open to enterprise and experiment on the part of local school boards and unofficial educational bodies. The most notable and long-lasting of these experiments took place in London, Ontario, where the Central Collegiate Institute, under its principal Mr. Everton A. Millar, pioneered by establishing in 1937 its own radio studios and producing therein its own school broadcasts. Subsequently these programmes were broadcast over facilities offered by the local station CFPL (London) to other schools in the London area. During the next five years, 137 programmes were put on the air, including two series dealing with the provinces of Canada and other parts of the British Empire. Recordings were also made off the air from broadcast speeches by Sir Winston Churchill, Franklin D. Roosevelt, King George VI and Queen Mary, and so on, which were rebroadcast to groups of students in the auditorium. Mr. Millar stressed the role played by teachers and students in building and running the studio and its equipment. His conclusion was that students would have an important part to play in the future production of educational broadcasts generally.

During the same period, interest continued to grow among other educators. Through the CBC and other sources, these educators were made aware of the successful development of school broadcasting in Britain and the United States. It was noticed that British school broadcasts had set a pattern for permanent and successful programmes in Nova Scotia and British Columbia. Recordings of BBC school broadcasts were demonstrated in Toronto and other Ontario centres, proving the high standard of educational and technical excellence they achieved.

CBS "SCHOOL OF THE AIR"

During 1940–41 the first CBS School of the Air broadcasts were heard in Ontario, over CBL Toronto, CBO Ottawa, and some private stations affiliated to the network in the province. Two courses, of music and literature ("Tales from Far and Near"), each of 26 half-hours, were broadcast. The programmes were found to have some general value as supplementary material in senior elementary grades and junior high school grades in Ontario, but were criticized for their lack of correlation with the Ontario course of studies and their lack of specific reference to Canada and its problems. The School of the Air broadcasts, therefore, tended to stimulate a greater demand in Ontario for the presentation of specific Canadian school broadcasts. Appreciation of the programmes came from certain boards of education, Home and School associations, children's libraries, the Junior Radio Course, and a number of individual teachers. In some areas, particularly in Toronto, the children's libraries made extensive provision to enable classes from adjacent public schools to visit them and hear the broadcasts of the children's literature series, "Tales from Far and Near." In July 1941 a full-scale demonstration of the School of the Air was given in Toronto at Hart House Theatre, with a simulated radio studio and a model class-room with pupils and teachers on a divided stage. Copies of the teachers' manual (125 pages) issued by the CBS, to cover the syllabus of the School of the Air broadcasts, were widely distributed to Ontario teachers and helped to focus educational attention on the possibilities of school broadcasting in Canada.

OEA INVESTIGATES SCHOOL BROADCASTING

In the fall of 1941 the Ontario Educational Association—a loosely knit but influential "parliament" of major educational organizations in the province—began to reflect this increased interest in school broadcasting. Following conversations between the Association's president, Mr. Lloyd White, and its Secretary, Mr. H. P. Sutton, on the one hand, and Mr. C. R. Delafield and the writer, both of the CBC, on the other, the policy committee of the OEA decided to set up a special 23-man committee to investigate the possibilities of school radio in Ontario. The chairman of this committee was Mr. W. B. Collier of Duke of York School, Toronto, and its secretary Mr. G. H. Dickenson, of Central High School of Commerce, Toronto. The membership of the committee was drawn largely from the Toronto area, but included teachers from York township, New Toronto, and Hamilton, as well as the editor of *The School* magazine (Ontario College of Education), and a representative of the Ontario Department of Health. Mr. Delafield and the author attended the meetings in an advisory capacity, on behalf of the CBC. Two full meetings and three subcommittee meetings were held between January and March 1942, leading to the presentation of a report at the 1942 annual convention of the Association.

The summary and recommendations of the report stressed ten points: school broadcasting had proved its worth in Britain and the United States and in six Canadian provinces; Ontario was lagging behind the rest of the country; British Columbia and Nova Scotia had equipped many schools with receivers; Ontario should take advantage of the CBC's expressed willingness to help; CBC programmes contained many broadcasts of general educational value, which should be publicized among teachers and students by the OEA and its members; Ontario teachers should not be left dependent for school broadcasts solely on American offerings such as CBS School of the Air and NBC's Damrosch Music Hour; school broadcasts must not supplant, but supplement, class-room teaching; music, science, English, and social studies were the most suitable subjects for school broadcasts, which should be aimed at students in grades up to 10; education for citizenship should be stressed in school broadcasts; and in due course a provincial provision for school broadcasts, supported by the Department of Education, should become feasible.

At Easter 1942 this report was endorsed by the Ontario Educational Association which at the same time reappointed its special committee with the duty of continuing to work for school broadcasting. The outcome was the holding on April 9 of a public demonstration and conference under the joint auspices of the OEA and the CBC. The conference took place at Eaton Auditorium in Toronto, and was attended by approximately 600 teachers and educators. The chair was taken by the writer, who introduced two eminent American educators, Dr. Willard Givens, Executive Secretary of the National Education Association, and Dr. Lyman Bryson, CBS Director of Education. Both gave addresses on "Radio and Education." Two demonstration broadcasts were afterwards given of programmes from the CBS School of the Air of the Americas, one of which had been contributed by CBC.

The resolutions passed by the OEA, the Ontario Federation of Home and School, and various meetings in favour of school broadcasting were reinforced in August 1942 by a resolution passed at the annual convention of the Canadian Teachers' Federation, and by another resolution passed by the Board of Education of the City of Toronto, both favouring school broadcasts. They were not without effect upon the attitude of the Ontario Department of Education.

ONTARIO CONTRIBUTES TO NATIONAL SERIES

During the summer of 1942 the CBC addressed a circular letter to all departments of education and the Canadian Teachers' Federation, inviting them to co-operate with it in presenting, over the CBC network, an experimental series of national school broadcasts entitled "Heroes of Canada," dramatizing the lives of Canadians who had contributed to Canada's growth by displaying the pioneering spirit and the sense of social responsibility. All the departments except Quebec agreed to support the project. The Ontario Department contributed to the series two twenty-minute programmes for which it assumed responsibility for the cost of script, acting, and musical talent. The subjects chosen by the Department for its two presentations were "With Axe and Flail—a Story of United Empire Loyalists" and "With

Pack and Pick—a Story of Northern Ontario." These broadcasts went on the air on November 6, 1942, and February 5, 1943, and were well received by teachers and students in Ontario schools.

EXPERIMENTAL MUSIC PROGRAMMES

A further experiment in Ontario was undertaken during the same winter, dealing with the subject of music appreciation. The CBC, at the request of Home and School representatives, teachers, and musicians, presented on Wednesday afternoons a series of ten 45-minute programmes entitled "Music for Young Folk," with Sir Ernest Mac-Millan, the Toronto Symphony Orchestra, and a number of vocal and instrumental soloists. The broadcasts were heard over a network of southern Ontario stations. A sixteen-page booklet containing illustrations of the instruments of the orchestra was prepared and issued by the CBC to accompany this series. Five thousand copies of the booklet were distributed to music teachers, with the co-operation of the Department of Education, through the Provincial Supervisor of Music, Dr. G. Roy Fenwick.

After the broadcasts Dr. Fenwick sent out a questionnaire to music teachers in the province, and received 240 replies, showing that 156 classes or groups of students had heard all or part of the series. The main reason why all the programmes were not heard was lack of satisfactory receiving equipment. Many suggestions for improvement of the programmes were made, such as more participation by children, greater use of familiar songs and music selections, greater use of choral and vocal music and less of instrumental music, inclusion of commentary by an experienced school musician rather than a distinguished academic musician, and earlier supply of programme information. Many of these suggestions were adopted by Dr. Fenwick and embodied in subsequent Ontario music broadcasts. He became keenly interested in the possibility of developing a wider and more systematic use of broadcasting in music education in Ontario schools.

Ottawa, too, caught the infection of interest in school broadcasting. During February and March 1943, concurrently with the latter part of the national series referred to above, an experimental series of local school broadcasts was presented over station CBO (Ottawa). The programmes, under the title "Trumpet Call to Youth," dramatized the contributions made by new Canadians to Canada's national culture and growth. Groups of local public and high school students shared in the presentation of the broadcasts, which were also linked to a series of film showings covering the same topics, presented at different times, by the National Film Board. Notes for teachers were circulated, and posters prepared by the art classes in Ottawa schools. French-speaking schools in the Ottawa district voiced the hope that similar programmes might be broadcast in French for French-speaking schools in Ontario.

In the city of London (Ontario) the Central Collegiate Institute continued to broadcast, over station CFPL, programmes prepared by students and teachers in the Institute.

Throughout the winter of 1942, Mr. Delafield and the author, on behalf of the CBC, visited personally a number of schools in Toronto and district, in order to be present in the class-rooms while the school broadcasts were being received. Thereby they gained valuable experience, which proved useful in improving future programmes.

VIEWS OF LISTENERS

Many appreciative letters about the programmes reached the CBC from individual parents, from children, and from adult listeners generally. Typical of these was a letter from the mother of a family living in a lighthouse on Lake Superior, whose children, cut off by distance and illness from attending public school, were getting some of the elements of their education through the radio. School children themselves wrote many letters praising the broadcasts. One boy wrote on behalf of his class asking for a set of report forms, in order that he and his fellow students might send in their own views and criticisms, instead of leaving it to the teacher. Another gave his opinion on the Ontario contribution to the national school series: "I liked 'With Axe and Flail' because it was exciting and interesting, especially the part where the tree nearly hit the child, but injured the man who tried to save it. I learned a lot where the people ran into the water for safety from the fire." A third student commented on "Music for Young Folk": "I liked the programmes with the individual instruments. I also liked the programmes of music that told stories. Some of the music was hard to understand—such as Norwegian pieces and some from Central Europe."

Naturally, the growing public awareness of school broadcasts produced a flood of requests to address meetings of teachers, parents, and school trustees in all parts of southern Ontario. At this time there was no one in a position to meet these requests except the representatives of the CBC. Accordingly, the author addressed the annual convention of the Ontario Urban Trustees at Chatham, Ontario; the music and audio-visual sections of the Ontario Federation of Home and School at its annual convention in Toronto; the Toronto Teachers' Council; and various meetings in York township, Hamilton, and other centres. From the CBC also were sent out numerous reports, memoranda, and copies of scripts to teachers and educators in all parts of the province. At the same time articles on school broadcasting appeared in the press. For example, *Toronto Saturday Night, The School* magazine, the *Ontario Public School Argus*, the *Bulletin* of the Canadian Home and School Federation, the Ontario *Home and School Review*, the *St. Joseph Lilies, School Progress, Canadian Junior Red Cross*, and similar publications. The cumulative effect of all this publicity was to create a strong demand for a regular series of Ontario school broadcasts.

A steady increase took place in the number of radio receivers installed in Ontario schools. In its 1943–44 estimates, the Toronto Board of Education voted a sum of money for providing receiving equipment for all the schools in its area. In Ontario as a whole, the Department of Education estimated that some one thousand schools of all types, including 680 public schools, were now equipped with receivers. Admittedly, many of these receivers—especially in rural schools—were of the small mantel type; on the other hand, a number of larger schools were now equipped with public address systems, and a few school buildings were equipped with complete sound systems.

EFFECTS OF POLITICAL CHANGES

During 1943 changes took place in the political situation in Ontario. In August of that year a provincial election was held, leading to the fall of the Liberal government, and its replacement by a Progressive Conservative administration under Colonel George Drew. The new Premier, who took a keen interest in education and culture, undertook to combine the portfolio of education with the premiership. As Minister of Education, he was responsible for many progressive innovations in the Ontario school system, one of the first being the provision of funds for school broadcasts, planned and presented by the Department of Education in co-operation with the CBC.

Here we may properly pay a tribute to Fiorenza Drew, the Premier's wife, for her keen interest in the contribution that radio could make to music appreciation among school children. As an influential member of the women's committee of the Toronto Symphony Orchestra, Mrs. Drew was in a position to convey to her husband the committee's wish that experiment should be encouraged in this field of school broadcasting. Accordingly, in the summer of 1943, the Department of Edution approached the CBC with a plan for joint presentation of a series of music appreciation broadcasts, along similar lines to that of the previous winter's experimental series "Music for Young Folk." Ten thirty-minute programmes were proposed, "designed to help children appreciate the beauty and significance of the world's great music." Dr. Fenwick was to act as commentator for the series, and a pamphlet outlining the selections to be performed was to be issued to schools by the Department. The CBC at once accepted the Department's proposal, and a joint committee was set up to undertake the planning of the details.

At the first meeting of the committee, Dr. Fenwick explained the nature of the Department's approach to music appreciation through radio. Rejecting, on the one hand, any attempt to *teach* pupils how to appreciate music or, on the other hand, the giving of *instruction* in music techniques (such as playing a musical instrument), he voiced his faith in the desirability of exposing the pupils to good music, in the belief that this would arouse their interest and cultivate in them an enlightened musical taste. By himself taking part in the broadcasts, Dr. Fenwick hoped to strengthen the personal link that already existed between himself and the pupils in the music class-rooms of the province.

Associated with Dr. Fenwick on the joint planning committee was Dr. Leslie Bell, of the Ontario College of Education, who undertook to prepare the scripts of the broadcasts. For this Dr. Bell was specially fitted by his wide knowledge of music, and by his experience as choirmaster and musical commentator. The CBC was represented by the author, then its Supervisor of School Broadcasts, and by one of its senior music producers, Mr. George Dixon. The fifth member of the committee was Mr. Paul Scherman, who undertook to assemble and direct the orchestra.

"MUSIC FOR YOUNG FOLK"

The first of the ten programmes, entitled "The Men Who Write Music," was broadcast on Friday afternoon, January 14, from 2:00 to 2:30 P.M. It was aimed at students in grades 7 and 8, and was produced by George Dixon in the studios of CBL (Toronto). The programme was introduced in person by the Hon. George A. Drew, Premier and Minister of Education, who stressed the importance of music in education in the following terms: "The Ontario Government wants to make sure that each of you receives an education which not only prepares you to earn a living, but also teaches you to enjoy life itself. To get the best out of life it is necessary to have some understanding of the arts, and of these one of the greatest sources of enjoyment is an appreciation of music. At a time when there is so much ugliness, cruelty and unhappiness in the world, it is more important than ever before that our children should grow up with a love of beauty and the better things of life."

Five musical items chosen to form part of this first programme were: Schubert's Entr'acte to "Rosamunde" and "Moment Musicale," Tchaikovsky's "Chanson Triste," and Brahms' "Waltz in A-Flat" and "Hungarian Dance No. 1."

On this occasion, and for every year thereafter until 1959–60, Dr. Fenwick acted as commentator. He rarely missed appearing in person in the studio, though sometimes the exigencies of his departmental work forced him to pre-record his comments. His fatherly and intimate style of speaking made him specially popular with junior pupils in the class-rooms of the province.

The remainder of the series "Music in the Making" comprised the following titles: January 21, "Music that Tells a Story"; January 28, "Music for Music's Sake"; February 4, "Recognizing a Country by its Music"; February 11, "Music Changes with the Years"; February 18, "Elements of Music"; February 25, "The Background of Music"; March 3, "Developing an Idea"; March 10, "Everyone a Performer"; and March 17, "Which Ones do you Hear?"

PROGRAMME SCHEDULE EXTENDED AND EXPANDED

Along with "Music for Young Folk," the Department of Education presented, over the CBC Ontario network, the national school broadcasts and three of the courses of CBS School of the Air. Altogether, then, there were available to Ontario schools in 1943–44, for the first time, five weekly series of school broadcasts of which two were of Canadian origin and three were of American origin. Two series were presented on Fridays, from 10:00 to 10:30 A.M. and 2:00 to 2:30 P.M. The other three were offered on Mondays (science), Wednesdays (geography) and Thursdays (literature) from 10:00 to 10:30 A.M. All these programmes were also broadcast, at the special request of the Quebec (Protestant) Department of Education, over CBM (Montreal) at the same periods.

In March 1944, the newly formed National Advisory Council on

School Broadcasting held its first meeting in Toronto. On that occasion the Ontario Department of Education was ably represented by Dr. C. F. Cannon, the Superintendent of Elementary Education, who reported to the Council on the success of "Music for Young Folk." Dr. Cannon, who continued to sit on the Council until 1946, played an influential part not only on the Council itself, but also inside the Department in developing the policy, initiated by Premier Drew, of gradually expanding Ontario's provision of school broadcasts.

During 1944–45 the Department's programmes were all concentrated on Tuesdays, using two periods, from 10:00 to 10:30 A.M. and from 2:00 to 2:15 P.M. Before Christmas two series were given, social studies ("Pioneer Days") for grades 6 to 8, and guidance ("Vocational Guidance") for grades 9 to 13, in the morning period; and one, literature ("Junior Stories") for grades 1 to 4 in the early afternoon period. After Christmas, both periods were given up to music appreciation, with "Music for Young Folk" (grades 7 and 8) in the morning, and "Junior School Music" (grades 3 to 6) in the afternoon.

In the "Junior School Music" series, Dr. Fenwick presented songs familiar to the schools, introduced by himself, and performed by professional singers.[1] Dr. Fenwick's aim, in which he succeeded, was to help students, particularly in rural schools, learn the songs by participation in their own class-rooms during the broadcasts. The distinguished pianist, Mr. Leo Barkin, acted as accompanist, a role he has continued to fill with great success down to the present time. In later years, Mr. Barkin sometimes included solo performances.

In social studies, the Department co-operated with the Royal Ontario Museum and the CBC to present a series of five half-hour broadcasts on pioneer life in the province, aimed at grades 6 to 8. The programmes, entitled "The Saga of the Pioneers," were each divided into a ten-minute talk dealing with the physical and scientific factors influencing pioneer settlements, and a twenty-minute dramatization dealing with the social and economic life of the pioneers themselves.

Another series consisted of playlets for adolescent boys and girls on the subject of "guidance," introduced and directed by Mr. H. R.

[1]In the early days of this series, Dr. Fenwick's role as commentator did not go uncriticized by the CBC. It was felt by Mr. E. L. Bushnell, then CBC Programme Director, that Dr. Fenwick's voice and delivery did not come up to the standard expected from professional announcers and commentators on the national radio. However, when, as a result of visits of inspection to various schools, the popularity of Dr. Fenwick's voice and manner with the primary and junior grades of pupils had been demonstrated to him, he withdrew his technical objections, and allowed that, *for specific purposes*, an "amateur" voice might be more suitable than a "professional."

Beattie, who at that time was Ontario's Director of Guidance. Mr. Beattie had a natural gift for broadcasting, and his interpretations of the dramatic portions of the scripts were greatly appreciated in the class-room. He followed the Department's policy of emphasizing the "personal" rather than the "vocational" aspects of guidance.

All these series indicated that Ontario was anxious to develop effective presentation of school subjects through the use of dramatizations with "live" actors and musicians rather than by lesson-talks or recordings. Also, Ontario preferred to group its programmes in short series of from five to ten broadcasts rather than in long series of twenty-six to thirty broadcasts. This meant of course, spending more money on the production, and more time on the organization of each individual programme. Because not only outstanding teachers, but also professional writers, actors, and musicians were employed, the cost of the broadcasts was high. From the outset the Department, on the initiative of Dr. Cannon, showed itself to be quite prepared to allocate an adequate and even generous budget to the development of the school broadcasting work. This policy soon began to show its effect in the high quality of the broadcasts themselves.

APPOINTMENT OF MAJOR JAMES GRIMMON

It quickly became apparent that the expansion of the work would require the services of a full-time departmental official. Dr. Cannon decided that, because school radio was one of a group of new classroom "aids to teaching," it might usefully be combined with films, film strips, and other audio-visual aids under a single new department. Accordingly, in September 1945 the Department invited Major James W. Grimmon to become its Director of Audio-Visual Education. Major Grimmon, an Honours graduate of Queen's University, Kingston, and B.Paed. of the University of Toronto, had taught elementary school in Belleville and Port Credit, and in 1938 was appointed Principal of Elora High School. During World War II he served in the Canadian Army with the rank of Major, first as District Army Examiner and subsequently as Director of Personnel Selection at Military Headquarters in Toronto. In September 1945, at the special request of Premier Drew, he was discharged from the Army to take up his new appointment.

Mr. Grimmon soon showed himself a skilful diplomat and a capable organizer. However, the demanding nature of his job and the large number of schools in the province made it impossible for him to take personal charge of all the details of school broadcasts as was done in the four Western provinces. From the beginning, therefore, it was necessary to devise machinery whereby the departmental officials and the officers of the CBC School Broadcasts Department could co-operate closely by dividing the burden of the work between them.

A reasonable working solution to this problem was soon found. The Department kept closely in its hands the planning of school broadcasts, the allocation of the production budget, the responsibility for all scripts, and most of the class-room utilization, and, on the other hand, the CBC added to its staff a producer and a programme organizer who specialized in the presentation of the Ontario school broadcasts. The writer shared with Mr. Grimmon some of the promotional work among Ontario schools. He and his staff also answered a good deal of the correspondence from teachers, parents, and students on matters arising out of the individual broadcasts.

ONTARIO TEACHERS' MANUAL

Mr. Grimmon's first task was to produce a manual for the teachers of Ontario, giving them the necessary information about the coming programme of Ontario school broadcasts for 1945–46. Copy for it had to be rushed to the printers within forty-eight hours of his arrival on the scene and so quick decisions had to be taken.

The manual had to be distributed over a province which comprised roughly one-third of all the schools and teachers in the whole of Canada! To publish a full manual of, say, 100 or more pages (such as was done in Manitoba and other provinces), with illustrations, would be very costly. Including mailing charges, it would use up a major share of the funds allocated to school broadcasting from the departmental budget. Mr. Grimmon decided it would be preferable to spend more money on the programmes themselves and less on the manual. Accordingly, the manual that year and the next was limited to eight pages leaving room for little more than titles and dates of the individual broadcasts, brief notes on each series, and an introductory foreword. In the later issues slightly more detail, including lists of recommended films and film strips, was added.

From time to time, the Department received complaints from teachers about the inadequacy of the manual. Teachers declared that they needed much more information about the broadcasts to ensure their effective utilization. In the 1950–51 manual, for the first time, these needs were met in the case of one series, Miss McCance's "Adventures in Speech," by the inclusion of the text of her "speech exercises." In 1953 the manual was divided into two separate sections, one for elementary, the other for secondary school teachers. The elementary section was standardized at 28 pages, the secondary at 12 pages. In 1960, 40,000 copies of the elementary manual and 11,500 copies of the secondary manual were printed. A smaller separate manual was also issued, from the beginning, for the music broadcasts.

SUMMER COURSE

In the summer of 1946, Mr. Grimmon started, at the Royal Ontario Museum, the first summer course in audio-visual aids for Ontario teachers. Though the course dealt primarily with film and similar visual aids, a section of it was devoted to school broadcasting, including the giving of information and instruction and opportunities for practising utilization and planning of school programmes. Later, speakers from the CBC assisted by giving addresses on national and provincial school broadcasting. In 1946 only 33 students attended the course, but by 1960 the number had grown to 300. Two other courses of the same type were added by the Department in outlying centres of the province.

DEPARTMENTAL RADIO COMMITTEE

For the planning of the Ontario school broadcasts, a Departmental Radio Committee was set up by the Minister of Education. Its membership, which varied from five to nine, included the Superintendent of Elementary and Secondary Education, the Registrar, the Superintendent of Professional Training, and representatives (especially for music and guidance) of the elementary and secondary inspectorate of the province. Dr. C. F. Cannon acted as chairman for the first three years; he was succeeded by Mr. S. Holmes, who continued till 1959. Mr. J. W. Grimmon served as Secretary, and in 1959 succeeded Mr. Holmes as Chairman. The Radio Committee held frequent informal meetings, and one formal annual meeting at the close of each school broadcasting year at which the plans for the following year's broadcasts were discussed and adopted. No outside bodies, such as the CBC or the Ontario Teachers' Federation, were invited to be present on these occasions.

PLANNING AND PRODUCTION

As soon as the Radio Committee had decided on its plans, an estimate of their cost was prepared by Mr. Grimmon in consultation with the author. When the Minister had approved this cost, the budget for school broadcasts was finalized, and it became the responsibility of the author to see that the programmes were produced in accordance with the sums allocated. He recommended to Mr. Grimmon suitable script-writers, who were subsequently commissioned and paid by the Department. For each series (usually five to six programmes) the Department appointed a paid consultant, usually an inspector or a teacher who specialized in the subject chosen. The consultant, together with the CBC programme organizer and the producer assigned to the job, acted as a subcommittee responsible for carrying the series into execution. In all cases drafts of scripts were submitted to the Department for final approval before going on the air. This machinery worked very smoothly, and only rarely did any complaint or difficulty have to be referred to the ministerial level.

NOTABLE PROGRAMME SERIES

Ontario has benefited from the presence of the CBC English-language programme headquarters in Toronto, for the best possible talent in the script, acting, and music fields, as well as the best specialized production and technical talent, has been available for the Ontario school broadcasts. Hence it is not surprising that social studies programmes, in particular, have been strong in historical dramatization, while the music appreciation programmes have been enriched by the regular employment of "live" musical talent, orchestral, instrumental, and vocal. Because of the rich professional talent available, fewer teachers have been used at the microphone than in other provinces. Even here, however, there have been notable exceptions when individual teachers were discovered who possessed the natural gift of "radio personality."

Health Broadcasts

One of the earliest successful broadcasters was Mr. Fred L. Bartlett, Provincial Director of Physical and Health Education. Mr. Bartlett commenced broadcasting in November 1945, and continued his programmes during 1946 and 1947. Each of his half-hour broadcasts was divided into two parts. One was occupied by a dramatization of factors contributing to good health. The other was an "activity" programme in which pupils participated in their own class-rooms, as directed over the radio. Mr. Bartlett had the rare gift of being able, in the studio, to visualize his audience in the class-rooms of the province. It was a striking experience to visit one of these class-rooms, and see the students responding to the unseen voice that directed their physical movements. They would stretch and wave their limbs, climb on their desks, march about the room—all in accordance with the staccato commands of Fred Bartlett, and all obviously enjoying what they were doing. In 1947 Mr. Bartlett left the Department of Education to become head of the Physical Education Department of Queen's University, Kingston. Later, he became President of the Ontario School Trustees' Association. After his departure, no one was found able to replace him, and the programme lapsed.

Junior Stories

A regular feature of the Ontario programmes for over fifteen years was "Junior Story Period." The stories were given in narrated form to children of kindergarten and grades 1 and 2, and in dramatized form to children of grades 3 and 4. In the early days the narrated junior stories were chosen by Miss Lillian Smith, the late Head of Boys and Girls House (Toronto Public Libraries), and often performed by members of her staff. The dramatized stories were scripted by Miss Dorothy Jane Goulding (Mrs. W. Needles).

In general, Miss Smith belonged to the school of thought that favoured giving very young children the traditional fairy stories and legends in their time-honoured form. She had little sympathy with modernized versions of these tales, still less with contemporary literary creations. However, after the passage of several years, a widening of the scope of the programme took place. At the same time, it was found, from teachers' reports to the Department, that children generally favoured the "dramatized" over the "narrated" stories. Latterly, all programmes in the series were presented in dramatized form.

Several amusing incidents occurred in connection with the junior stories. "Little Red Riding-Hood" used to figure regularly in the series during the early years, but in a slightly expurgated form. The late John Walsh, Inspector of Schools, who supervised the scripts on behalf of the Department, was greatly upset to find that in the traditional form of the tale Little Red Riding-Hood carried "cakes and a bottle of wine" in her basket when she visited her grandmother, and he insisted that "strawberry jam" be substituted for the wine. In October 1948 "Little Black Sambo" by Helen Bannerman figured in the story period without arousing any particular comment. But when it was repeated four years later, a considerable outcry was raised that the mere title (not the story) was calculated to encourage racial prejudice. The story had to be dropped from the repertoire.

On various occasions parents and psychologists who listened to these stories took up conflicting attitudes towards the "horror element" in traditional fairy tales. Some mothers complained that their small children's sleep was disturbed by the impression of violence and cruelty received from hearing Grimm's versions. Yet when the text of the stories was suitably softened to meet the objection, others raised their voice to protest that the emasculation would tend to produce a "soft" younger generation!

Literature

At the junior level, the aim of such broadcasts was to encourage young people to read more good books. Each year a selection was made of the best of recent children's books. Episodes from these books were dramatized, and often personal interviews with the authors included. Librarians often testified that such broadcasts were followed by a spate of requests for the books dealt with.

Social Studies

Ontario paid particular attention to social studies in the fields of both history and geography. Dramatized broadcasts (at first of fifteen minutes', latterly of thirty minutes' duration) on Canadian explorers and on aspects of Canadian, British, and American history were frequent. The "cycle" system was introduced, whereby social studies broadcasts for grades 4, 5, 6, and 7 were repeated every third or fourth year, by which time the student audience would have changed. In geography, "actuality" broadcasts in the form of visits to industrial centres were often given with success.

"Current Events" proved to be one of the subjects most helpful in the class-room, and was given weekly for many years. Mr. Vick Dobson, an Ontario teacher who served as Assistant Supervisor of School Broadcasts for twelve years, conducted this programme with great success.

High School Programmes

Ontario was one of the few provinces that made regular provision of broadcasts to high schools, usually at the grade 9–10 level, but sometimes at the more senior level. English literature, especially as related to the works specified in the course of studies, was the most acceptable. These broadcasts took the form of talks by outstanding teachers, often illustrated by dramatized excerpts from books. Scenes from Shakespeare were often included. In the field of history, dramatizations were given of episodes in mediaeval and modern European history, one of the most notable of which was a fully dramatized series on the French Revolution. Regular broadcasts on Greek and Roman classics were included for many years. Sometimes these dealt with the relations between the Latin and English languages, while at other times famous episodes in ancient history, or episodes illustrating Roman social life were dramatized.

All these high school series unfortunately had a relatively restricted audience appeal, as compared with the elementary school broadcasts, owing to the rigidity of the curriculum and time-table of high school teachers. However, many schools which acquired tape-recorders made off-the-air recordings of the programmes for further study and repetition in class.

Other Programme Features

In science, Mr. Grimmon took a keen personal interest in the provision of broadcasts dealing with Canadian wildlife and conservation. In 1948 he was responsible for the first series of Ontario broadcasts (grades 6, 7, and 8) on "Animals and Birds of Canada." This proved highly successful and bore fruit two years later in the starting, by the National Advisory Council, of a series "Voices of the Wild" (grades 3 and 4), which became famous and popular in the schools of Canada as a whole. At a later date, the national series was repeated in the Ontario schedule, in order to give Ontario schools both a spring and an autumn series of this kind.

Ontario, with its high respect for the British tradition in education, was the province which made most regular use of the selected series offered each year by the BBC, on transcription from its domestic school broadcasts. Ontario also welcomed the arrangements made by the school broadcast departments of the CBC and the Australian Broadcasting Commission, and made regular use every year of the resulting programmes, and of the inter-Commonwealth school broadcast exchanges which developed from 1949 onwards.

Ontario also actively interested itself in programmes presented in the schedules of other provincial departments of education, and included these, by arrangement, in its own schedule. Among these the most notable was the Manitoba series "Adventures in Speech," and the Maritime series of science talks (for grades 7 and 8) given by Dr. Lloyd Shaw, at that time Deputy Minister of Education for Prince Edward Island.

ONTARIO STATION NETWORK

A great deal of the success of the school broadcasting work in the province has been due to the extensive network of stations established by the CBC to carry the programmes. Basic CBC stations at Toronto, Ottawa, and Windsor have been supplemented by privately owned stations in all parts of the province. In addition, from 1945 onwards, at the special request of the Quebec (Protestant) Department of Education, the Ontario school broadcasts were made available to English-language stations in Quebec, including CBM Montreal and CJQC Quebec. The original "mid-Eastern" network has thus been expanded from 15 stations in 1944 to 27 in 1960, without counting CBL and CBM repeater stations.

After 1949 the Department of Education decided to present all its school broadcasts in a morning period from 9:45 to 10:15 A.M. EST, which at that time seemed to fit in best with class-room requirements. However, as time passed, individual private stations carrying the series asked permission, mainly for commercial reasons, to delay the programmes to the afternoon, with the consent of the education authorities. By 1960 the majority of stations had obtained this consent, and were delaying the broadcasts accordingly. By this time also it had become apparent that most teachers in the province preferred an afternoon period. In 1957 the Department secured CBC's approval for a change to a later morning period, 11:00–11:30 A.M., in the hope that at a later date an afternoon period might become possible.

RADIO-COLLÈGE

When the National Advisory Council on School Broadcasting was set up in 1942 its constitution provided for the appointment, by the Quebec Department of Education, of two representatives, one for the Catholic, the other for the Protestant sections of the Department. The Liberal government then in office in Quebec appointed the two representatives, Dr. B. O. Filteau (Catholic) and Dr. W. P. Percival (Protestant), but it did not accept the CBC's invitation to use its facilities for any provincial school broadcasts or participate in the provision of national school broadcasts. Quebec laid heavy stress upon its constitutional prerogative in all matters of education and, regarding the CBC as essentially a federal agency, showed no desire to take advantage of whatever facilities the CBC offered to the Department of Education. The Protestant (English-speaking) section, however, was permitted to ask the CBC and the Ontario Department of Education to arrange that the Ontario and national school broadcasts should be heard over CBM and other English-language stations in the province for the benefit of Protestant schools.

Quebec Refuses Co-operation with CBC

When, after the War, the Liberal government in Quebec was displaced by the Union Nationale, Premier Maurice Duplessis planned to establish a complete network of stations owned by the provincial government, over which programmes of which he approved (such as school broadcasts) might be presented. However, his plan came to nothing for, according to the Canadian constitution broadcasting is a function of the federal rather than the provincial government.

When Dr. Filteau retired from the service of the Quebec Department of Education in 1954, he was not replaced, and from then on the only link between the Department and the National Advisory Council was through the representative of the Protestant section.

As early as 1939, Dr. Augustin Frigon, who as Assistant General Manager of the CBC had special responsibility for the network of French-speaking stations in Quebec, interested himself in the possibility of doing something to fill the educational gap created by the Department of Education's abstention from school broadcasting. He asked Mr. Aurèle Séguin, a former teacher who was assisting the manager of CBV Quebec in connection with the Royal Visit to Canada that summer, to prepare a plan of educational programmes for listeners in the province.

Dr. Frigon Founds Radio-Collège

The lack of co-operation of the Department of Education ruled out the possibility of broadcasting programmes to elementary schools on the same lines as the national and provincial school broadcasts in English-speaking Canada. However, the secondary schools of Quebec, especially the collèges classiques, did not come directly under the Department of Education, but under the universities of Laval and Montreal. Mr. Séguin therefore recommended the provision, under the name of Radio-Collège, of a series of educational programmes which could serve as a general supplement to high school studies and at the same time be of interest to adult listeners who wished to complete their own education. He asked Dr. Frigon for fifteen such programmes a week, to be broadcast in the late afternoon between the hours of 4:30 and 5:30 P.M. Altogether there would be three and three-quarter hours of such broadcasts, heard in out-of-school periods—considerably more time per week than was available for school broadcasts on the English network.

Dr. Frigon accepted the project outlined by Mr. Séguin, and allocated to it a budget of $5,200 a year, excluding the cost of a 36-page programme booklet of which 5,000 copies were printed for distribution to high schools and adults. In a speech at the opening of *Radio-Collège* in 1941, Dr. Frigon voiced the general aims of the new institution as follows: "If our young people, our professors, our teachers, our school inspectors, our schools, our listeners generally derive some advantage from *Radio-Collège*, we shall be quite satisfied and we shall have played the part which we assigned to ourselves. We hope that the broadcasts of *Radio-Collège* will be an encouragement to them, and a help in carrying out their educational work or in the perfecting of their own culture."

Programme of Radio-Collège

To plan the programmes, a standing committee of advisers was set up by Mr. Séguin, consisting of the following well-known educators: Abbé Georges Perras, President of the Standing Committee on Secondary Education, University of Montreal; Abbé Emile Beaudry, President of the Standing Committee on Secondary Education, Laval University; Rev. Father Alcantara Dion, Secretary of the Standing Committee on Secondary Education, Laval University; and Rev. Brother Marie-Victorin, Director of the Botanical Gardens and of the Botanical Institute of Montreal University. The secretary of this committee was Mr. Aurèle Séguin who had by then been appointed Director of Educational Broadcasts, CBC French Network.

The first winter's programme was as follows:

Monday: 4:30–4:45 P.M., "The Laws of Nature," by Professor Léon Lortie; 4:45–5:00 P.M., "History and Application of Science," by Professor Louis Bourgoin.
Tuesday: 4:30–4:45 P.M., "History of Canada," by L'Abbé Albert Tessier; 4:45–5:00 P.M., dramatized historical sketch; 5:00–5:15 P.M., natural history, "The City of Plants," by Brother Marie-Victorin.
Wednesday: 4:30–4:45 P.M., the arts–"Sculpture," by Gerard Morrisset, and "Architecture," by Jules Bazin.
Thursday: 4:45–5:15 P.M., literature–"Poetry," by Jean Charles Bonenfant, and "Drama," by Luc Lacourcière.
Friday: 4:30–5:00 P.M., "Music," by Claude Champagne.

The broadcasts were heard on the eight stations of the French network of the CBC.

For its second season (1942–43) the programmes were somewhat modified and expanded. The art and architecture courses were replaced by courses on speech (by Georges Landreau), human geography (by Professor Raymond Tanghe), and actuality (*"Radio-Collège* on Tour," by Aurèle Séguin). Also, an important series of classical French dramas was introduced on Sunday evenings from 8:00 to 9:00 P.M. under the title "*Radio-Collège* Theatre." Plays by Corneille, Racine, Molière, Marivaux, Musset, Victor Hugo, Sardou, Rostand, and many others were performed. These drama performances proved to be, over the years, the most successful and popular feature of *Radio-Collège*.

In addition to these broadcasts in French, *Radio-Collège* undertook to contribute two programmes in English, on behalf of French-speaking Quebec, to the first series of national school broadcasts "Heroes of Canada." The subjects of these two broadcasts were Jeanne Mance (pioneer nurse of North America) and Sir George Etienne Cartier (statesman of Confederation).

Aids to Study Publications

The circulation of the programme booklet of *Radio-Collège* rose to 15,000 in 1944–45 and in the following ten years to 30,000 copies. Most of the programmes of *Radio-Collège* were presented in lecture form, and a demand arose to have these in print. Accordingly, arrangements were made to publish (through Editions Fides of Montreal) the most outstanding lecture-courses in book form, under the general title *Les Editions de Radio-Collège*. Each volume when published enjoyed a circulation of from three to five thousand copies. Various "aids to study" booklets were also issued to go with the broadcasts, notably a brochure on plant life and a series of maps of the world to illustrate the history course of Abbé Tessier.

Also in 1942, Mr. Séguin started holding "Le Concours de *Radio-Collège*," an annual display in the fall at Montreal Botanical Gardens, of work produced by listeners to the past season's broadcasts. The exhibits came mainly from youngsters who had prepared them during the summer vacation and included essays, collections of plants, fruits, herbs and flowers, and photographs. Contests in music appreciation were also held.

First Radio-Collège Conference

To mark the fifth anniversary of its inception, *Radio-Collège* held its first Conference (March 11–12, 1946) at the Cercle Universitaire de Montréal. It was attended by Dr. Augustin Frigon, Director-General of the CBC and founder of *Radio-Collège*; Mr. Aurèle Séguin, its first Director, and Mr. Gérard Lamarche, his assistant; several members of the *comité pédagogique* of *Radio-Collège*, including Mr. Georges Perras (University of Montreal) and Abbé Emile Beaudry

(Laval University); Mr. Leopold Houlé, of CBC Press and Information; Mr. Florent Forget, CBC producer; and fifteen of the lecturers who delivered the courses of *Radio-Collège*.

Discussion took place on the planning-organization and production of the programmes, and demonstrations were given of individual programmes as practical illustrations. Drama, Bible study, literature, science, and history were examined in turn, and the outlines of future programmes for 1946 and 1947 were approved.

Dr. Frigon, in a general survey of *Radio-Collège*, laid stress on the spirit of perfection which had animated all those who had contributed to its courses. He also emphasized the needs of the listening audience, which he said must form the basis of any such educational broadcasting project. Lastly, he spoke in glowing terms of CBC's achievement in establishing *Radio-Collège*. "*Radio-Collège*," he declared, "is the finest of the CBC's achievements in the Province of Quebec [Radio-Collège, c'est le plus bel effort de Radio-Canada pour la province de Québec]."

In a report on the first five years of *Radio-Collège*, Mr. Séguin pointed out that the amount of air time given to *Radio-Collège* had steadily increased from three hours and 35 minutes per week in 1941–42 to six hours and 30 minutes in 1945–46. In addition, UNESCO had in 1946 requested help from *Radio-Collège* in supplying educational broadcasts on transcription for use in Germany, Belgium, France, Denmark, Holland, and Czechoslovakia to ease the postwar teacher shortage in those countries.

In the course of this conference, Mr. Roger Clausse called attention to the fact that secondary schools in general seemed to remain impervious to the influence of radio, partly because of the rigidity of the curriculum, partly because of their resistance to new educational methods.

Towards Adult Education

In spite of the fact that from 1946 onwards the expansion of *Radio-Collège* continued (as is evidenced by the growth in circulation of the programme booklet to 32,000 copies each season), a change was noticeable in the constituency of its audience. As was stated by Mr. Raymond David (who in 1954 succeeded Mr. Gérard Lamarche as Director),

The aims of *Radio-Collège* have had to be progressively modified, in order to respond to the needs of its listeners. From having been originally a supplement to school studies, the service has come gradually to widen its scope in order to cater to all CBC listeners who are interested in the things of the mind. Programmes in science, art, literature, history and religion, now give listeners generally the kind of material which helps them to gain better psychological balance, improve their scientific knowledge, develop their tastes for art, music and books, clarify their spiritual life, achieve a better adjusment to the times they live in, and interest themselves in the affairs of foreign countries. In brief, it helps both to reinforce Quebec's traditional culture and to keep the minds of our people open to new aspects and influences of modern life.

The very success of *Radio-Collège* in becoming primarily an institution for adult education also pointed the way towards its own eventual disappearance. According to Mr. David, soon after the arrival of television in Canada, *Radio-Collège* began to feel adverse effects from its impact. The audience decreased to one-half. Many of the programmes (especially in the sciences) could be (and were) more effectively presented through the new medium. At the same time a shrinking budget made it difficult to raise the standard of the broadcasts.

Efforts made to co-ordinate *Radio-Collège* with the educational television programmes were unsuccessful. The barrier between the old and the new medium could not be broken down quickly enough. Finally in 1955 the decision was taken to end *Radio-Collège* as a separate institution, and merge its programmes in those of the Talks Department, which at the same time became the Adult Education and Public Affairs Department of the French network. The new department was asked to take charge of all educational and public affairs programmes produced on television. In this way staff with radio experience were at last enabled to contribute towards helping television to assume its full and effective responsibilities in the field of educational and public affairs programmes.

Thus was prolonged and amplified one of the most valuable achievements of the French network in the field of popular education. Unfortunately, the policy of non-co-operation followed by the Quebec Department of Education did not begin to loosen until after the change of government in 1960. At that time, at last, there appeared to be a prospect that school broadcasts might be presented in French to the French-speaking schools of Quebec. If so, the technical personnel and experience of school broadcasting built up by *Radio-Collège* between 1941 and 1955 will be considerably missed.

VI. Newfoundland School Broadcasts

In Newfoundland, school broadcasting encountered conditions very different from those in other parts of Canada. At the end of World War II, the Island had a population of approximately 350,000 (according to the 1951 census, the figure was 361,000), the majority of whom were scattered in sometimes isolated pockets (outports) around a rugged coastline of over 6,000 miles. In addition, Newfoundland also suffered from inadequate communications (by road and rail), particularly in winter-time. It had only one large city (St. John's) with two or three smaller towns on the east and west coasts and in central Newfoundland. However, the major portion of the population lived on the Avalon peninsula. The standard of living of most Newfoundlanders was low by comparison with other parts of Canada. The population was still largely dependent on the three main industries—the cod-fishery, pulp and paper, and mining. However, other industries were gradually being developed. Considering these circumstances, the broadcasting system was of vital importance to the people of Newfoundland. Because the Island was an independent region its programming was characterized by an unusual flexibility, which opened up opportunities for the educational uses of radio to a greater degree than was feasible in more heavily industrialized regions of North America.

Prior to 1949 no school programmes had been provided for the Island. However, as soon as Newfoundland joined Confederation the possibility of extending school broadcasting to the new province by using the facilities of the CBC network, which took over the stations formerly owned by the Newfoundland Broadcasting Corporation, came under consideration. Dr. G. A. Frecker, the then Secretary for Education in the Newfoundland Department of Education, attended the sixth meeting of the National Advisory Council on School Broadcasting in Toronto in March 1949 and reported on the situation. He indicated that his Department was anxious to co-operate with the rest of Canada, and that Newfoundland would welcome receiving both the national school broadcasts and "Kindergarten of the Air," as well as taking part in the Maritime regional school broadcasts. He said also that Newfoundland would like to originate programmes of its own, not only as a contribution to the Maritime school broadcasts but also to meet its own provincial needs.

Without delay, the Newfoundland Department of Education was admitted to membership of the National Advisory Council, and two special programmes (aimed at the grades 6–9 level) were included in the national school broadcasts for 1949–50, for the purpose of better informing Canadian students about the geography, history, and life of Newfoundland.The programmes, "From Cabot to Confederation" (October 7, 1949) and "Life in the New Province" (October 14, 1949), were prepared in consultation with Dr. Frecker, and were generally well received in all parts of the country.

During the same season, the regular series of Maritime school broadcasts were heard for the first time in Newfoundland, and in 1952–53 Newfoundland began contributing programmes of its own to the series. A series of broadcasts on health (entitled "Keeping Fit" for grades 4–8) originated in the CBC studios in St. John's, produced by the present regional presentation officer there, Mr. Dick O'Brien. Mr. O'Brien did an excellent job in production, often working under difficult conditions and late into the night. Indeed, Dick O'Brien can be called the pioneer producer of school broadcasts in Newfoundland. The series on health ran from November 1952 to February 1953, and was followed by a series entitled "Our History" from October 1953 to January 1954. In December 1953 Newfoundland made its first contribution to the national series, in the form of a programme of Christmas carols contributed by the schools of four religious denominations in St. John's.

Prior to 1950 the visual education services of the Newfoundland

Department of Education were administered together with adult education under one division known as the Division of Adult and Visual
Education. However, in April 1950 the Department of Education,
realizing the value of audio-visual aids, undertook to set up a separate
Division of Audio-Visual Education. Included in the responsibilities of
the new Division were the promotion and maintenance of a programme
of radio education. The first Director of the new Division was a former
teacher and school supervising inspector, Mr. Frank Kennedy. In
March 1951 Mr. Kennedy took his place for the first time as Newfoundland's representative at the eighth meeting of the National Advisory Council. He soon proved himself active and enterprising, throw-
ing himself wholeheartedly into ambitious plans for developing school
broadcasting in Newfoundland to a point where it would meet the
special and pressing needs of the schools of that province.

TEACHER SHORTAGE PROBLEMS

Following World War II Newfoundland, in common with other
parts of Canada, found itself facing a shortage of qualified teachers,
partly because of increased enrolment in schools necessitating the employment of more teachers, and partly because of the lack of adequate
training facilities for teachers and the failure to recruit sufficient qualified prospective teachers to meet the demand. To aggravate the situation further many qualified teachers left the profession after 1949 to
enter more lucrative positions in both the provincial and the federal
civil service. Besides this many schools were staffed by inexperienced
and inadequately trained teachers.

The Council of Education was concerned with this problem and
sought ways to overcome it. Among the solutions arrived at was the
development of a series of school radio programmes which would serve
not only as direct teaching lessons and supplements to the teacher's
class-room work but would also act as demonstration lessons for inadequately trained teachers. Acting on the recommendation of the
Council of Education, the provincial government therefore approved
an increased expenditure for radio education for the year 1954-55 to
enable the Department of Education to institute a series of experimental school broadcasts directly for Newfoundland schools.

At the eleventh meeting of the National Advisory Council in February 1954 Mr. Kennedy outlined an experimental plan he had prepared for the commencement of Newfoundland school broadcasts along
lines involving the use of radio for direct class-room teaching, in con-

junction with visual aids and other teaching media and in correlation with tuition by correspondence to be introduced in the near future. He also announced that the experiment had been endorsed by the Newfoundland government and that the CBC had approved the necessary time required to air the programmes.

EXPERIMENT IN NEWFOUNDLAND

The experiment involved a considerable extension of the time allocated by the CBC in Newfoundland to school broadcasting, the appointment of additional CBC staff to produce the new programmes, the equipping by the Department of Education of a selected number of schools with radio receivers so that these schools (the majority of which were one room) could participate in a controlled experiment, and collaboration between the Audio-Visual Division and the Curriculum Division of the Department of Education in correlating the proposed teaching broadcasts with the curriculum, and in securing the best teachers to prepare and deliver the broadcasts.

Before the experiment was launched a careful survey was made of the schools by the Department of Education with the various denominational superintendents of education selecting the schools to participate in the experiment. In the summer of 1954, the author, then Supervisor of School Broadcasts for the CBC, accompanied by Mr. Lusty, Organizer of School Broadcasts for the Atlantic region, visited Newfoundland, and with Mr. Galgay, regional manager of the CBC Newfoundland region, held consultations with the Council of Education. Visits to various parts of the island were also undertaken by departmental officials accompanied by officials of the CBC and the support of the leading clergy of the various denominations was secured.

At the meeting between CBC and departmental officials in St. John's in August 1954 the feasibility of starting the experiment with only one series of programmes—"Reading for Understanding"—was envisaged. However, since many teachers were eager for other programmes, a considerable number of which were already being prepared, and the Department had already voted a considerable budget for the purpose, it was decided to increase the number to five per week as soon as possible in the new year of 1955. Subsequently, the Department of Education in September 1954 seconded Mr. Frank Furey, a school supervisor who had recently returned from Fordham University, to co-ordinate the experiment, organize the scripts and teachers, evaluate the results, and concentrate upon school radio. A year later, in July 1955, Mr. Furey became Assistant Director of the Audio-Visual Division and at the end of 1959, when Mr. Kennedy became Superintendent of Education (Catholic), was appointed to succeed him as Director of the Division.

Meanwhile in the autumn of 1954 Mr. Paul O'Neill, formerly a professional actor and at that time an announcer–operator on the CBC staff at Corner Brook, was appointed Producer–Organizer of School Broadcasts at CBC, St. John's, and took over the task of producing Newfoundland school broadcasts as well as those contributed by Newfoundland to the Atlantic region. At a later date, as the work developed, an assistant producer was appointed to assist him.

The stated purpose of the experiment, according to the Department of Education, was to find out "(a) how far 'direct teaching' by school broadcasts could be as effective as it potentially appears to be, (b) what type of equipment was most suited to the needs of individual classes, particularly in one-room schools, (c) what degree of teacher control was necessary to ensure that each broadcast would be an orderly episode in the daily routine of the classroom, (d) what educational objectives could definitely be achieved by educational radio programmes, and what resources and annual expenditure would be required to develop the project." It was also hoped that the experiment would help to decide the optimum length of each broadcast, the number of programmes required, the rate at which directions could be given over the air while remaining meaningful to the students, how the air could be used so as to enable pupils to answer direct questions, and how to secure effective correlation of broadcast material with related audio-visual aids to teaching.

Naturally all this called for a greatly increased expenditure on the part of the Department of Education. The budget for school broadcasts which started in 1951–52 at nearly $5,000 had risen to $7,200 for the year 1953–54. In the following year, with the introduction of the new programmes, it more than tripled that of the preceding year, reaching $22,273. It continued to rise to a peak of $27,238 for 1956–57 and thereafter declined and levelled off at approximately $20,000. The high exenditure in 1954–55 and following years was due to the Department's undertaking to install radio receivers on low rental terms in one hundred schools, selected on a denominational basis. The choice of schools was related to the local shortage of qualified teachers, and to restrictions on reception of local broadcasting stations in certain areas. Most schools so equipped at first were one- or two-room schools.

In December 1954 Mr. Galgay, on behalf of the CBC, agreed that five quarter-hour periods each week should be allocated to the experiment, as follows: Mondays to Thursdays, 11:30–11:45 A.M., Fridays, 2:15–2:30 P.M. These times were, of course, additional to those already given to the Atlantic regional school broadcasts and to the national school broadcasts. Altogether, then, the schools of Newfoundland had available to them in 1955 ten quarter-hour and one-half hour periods which were given to programmes of an "enrichment" and supplementary character, while the other five quarter-hour periods were used for "direct teaching."

The subjects of the "direct teaching" broadcasts were limited, according to a report prepared by Mr. C. F. Furey, to those which "could most effectively serve the diversified backgrounds and the capabilities of the maximum number of students in our schools, particularly the educationally underprivileged in our poorly-staffed schools." The following subjects were therefore scheduled: for high schools—French (grade 9), general science (grades 9–11), and English literature (grade 11); for elementary schools—"Reading with Understanding" (to develop reading skills) and "Reading for Pleasure" (to enrich the reading programmes to pupils in grades 4–8).

Results of the Experiment

The only detailed report on the experiment was that compiled in April and May 1955 by Mr. C. F. Furey, who piloted the experimental series. Mr. Furey made a detailed evaluation of the programmes, through teacher–evaluation forms followed up by personal visits of inspection. Reports were received from 70–75 per cent of the teachers using radios, from which it appeared that the course on general science had the greatest listening audience, with English literature next. "Reading for Pleasure" had the greatest popularity at the elementary level.

A considerable part of the province was not covered by satisfactory reception efficiency, however, and the poor quality of reception in some places hindered full use of the broadcasts. Tests prepared by the Department of Education were, of course, given to classes that had been utilizing the broadcasts. The reports appeared to show conclusively that "facts can be assimilated and concepts and ideas developed quite effectively by school broadcasts."

Experience showed that there was a great need for the better informing of teachers about both the availability and the utilization of school broadcasts. The former required sending out, before the beginning of the school year, advance details about forthcoming programmes and preparing teaching guides to accompany each series. The latter called for giving class-room demonstrations showing how the radio lessons could be used most effectively daily.

The general conclusions of Mr. Furey's report, as submitted to the Department of Education were that the experiment was "a very definite success"; that the experimental broadcasts were not so much concerned with "enrichment" as with "fundamentals"; that in many places where several grades in the same class-room listened to the broadcasts, teachers were of the opinion that, in certain subjects, there was need for broadcasts in specific subjects to meet the separate needs of those grades; that manuals must provide the teachers with "a concise and lucid lesson plan" to enable them to integrate the broadcast lesson with their school routine; and that the experiment should be extended to schools in Labrador. (The following year the Rev. F. W. Peacock, Superintendent of the Moravian Missions in Labrador, translated some of the scripts into the Eskimo language and produced them on the Mission station at Nain, Labrador.) Broadcasting of the Newfoundland school broadcasts later came to Labrador through the facilities of the CBC station at Goose Bay in September 1960.

In May 1955 the author and Mr. Lusty again visited Newfoundland and met with officials of the Department of Education and of the CBC, Newfoundland region, to assess the results of the experimental series. As a result of statistics gleaned from the report on the experiment it was felt that a definite contribution had been made to class-room work through the use of school broadcasts. This, coupled with a demand from teachers for more programmes particularly on the high school level, led to a continuation of school broadcasting on an expanded scale. From September 1955 school broadcasting became an accepted part of the Newfoundland system of education.

Several new programmes were introduced on the high school level and two quarter-hour morning periods per day were reserved by the CBC for the presentation of local provincial broadcasts. Included in the new programmes were detailed teaching lessons of Shakespearean plays prescribed for close study in high school. Dramatized performances were also given in half-hour instalments of Shakespeare's *Twelfth Night* and *Julius Caesar*. The number of programmes of Newfoundland school broadcasts continued to grow until in the school year 1959–60 the total number reached a peak of 526 programmes.

At the present time a definite pattern has been established with a school broadcast year beginning on the third Monday in September and continuing until the last week in May, permitting 32 fifteen-minute programmes in each of ten different series. The times utilized are from 10:30–10:45 A.M. and 11:30–11:45 A.M. NST, Monday to Friday.

CORRESPONDENCE COURSES

When the idea of beginning Newfoundland school broadcasts was first introduced in 1954 it was the intention that when tuition by correspondence would be introduced there would be some correlation between the Audio-Visual Education Division and the Correspondence Division. The Division of Correspondence Tuition was set up in 1958 and in September of that year the first lessons were sent out to students in grade 9. These lessons had been prepared the previous year under the direction of the Director of Audio-Visual Education pending the appointment of a Director of Correspondence Tuition.

In view of the fact that there was to be correlation between the two divisions Mr. Kennedy met with the author and Mr. Galgay in Toronto in February 1958 following the annual meeting of the National Advisory Council and submitted plans for a further expansion of school broadcasting in Newfoundland. The plans called for the equipping of all schools (mostly one- and two-room schools), whose students would do correspondence courses, with radios on a low rental basis to enable them to follow the school broadcasts, the majority of which would be for grade 9 students. The plans also called for a further extension of the time already allocated by the CBC in Newfoundland to school broadcasting. Mr. Galgay agreed that the CBC would grant a further one hour per week of broadcast time especially for correspondence students. As a result, two half-hour periods per week were approved for Correspondence Reports, on Mondays and Fridays at 11:15 A.M. and 11:45 A.M. respectively. These periods were utilized by the Director of Correspondence Tuition to report to correspondence students and to clear up any difficulties which they encountered in their work. The following year, at the request of the Director of Correspondence Tuition, only two fifteen-minute periods per week were utilized.

Subsequently the time was relinquished for various reasons, including lack of time on the part of the Director of Correspondence Tuition to prepare broadcasts and the decreasing number of students who were availing themselves of tuition by correspondence. At this time the government of Newfoundland introduced a very generous system of scholarships and bursaries which enabled students from smaller schools to complete their high school education in larger schools. However, those students still doing correspondence courses have the advantage of participating in the regular school broadcasts.

PROBLEMS

The large number of programmes presented each school year threw a heavy burden on the CBC producer in St. John's and necessitated the appointment of an assistant school producer at the St. John's studios. The Division of Audio-Visual Education was also faced with the problem of securing adequate script-writers who would prepare suitable scripts for radio presentation. To ensure that scripts were of an acceptable standard each one had to be approved and evaluated by the directors of the Curriculum and Audio-Visual Education divisions as to accuracy of content, correlation with text and curriculum, and so forth, before being released to CBC for production. The script was then transmitted to the CBC for approval as to production qualities. It was then returned to the Department of Education for the required number of copies needed for production.

Because the majority of the Newfoundland programmes were direct teaching lessons based on the textbooks in current use and because of the large number of broadcasts produced since 1954 without any major change in the textbooks it was becoming increasingly difficult to find new material on which to base scripts. Consequently the Department of Education found it necessary to repeat programmes already produced on tape. No difficulty was experienced, for an understanding with the writers permitted repetition of script material for a fee of 50 per cent of the original fee for a first repeat and 25 per cent for each repeat thereafter. The fee per script was $25 whereas the fee paid individual actors for participation was $7.50. A similar scale for repeat use of programmes on tape applied to actors. However, in 1960 the actor's fee was raised from $7.50 to $10 per broadcast permitting the outright use of tapes without further payment.

PART TWO

The Formative Period in National School Broadcasts

VII. The Beginning
of National
School Broadcasts

Two factors were mainly responsible for the start of national school broadcasts in 1942. The first was the creation of the CBC in 1936 and the subsequent setting up of a national network giving effective coverage over most of Canada. The second was the quickened national self-consciousness which arose from Canada's participation in World War II, and from her emergence as a powerful economic and political force in the postwar world.

Stimulating the development of educational broadcasting in Canada has always been one of the main functions of the Canadian Broadcasting Corporation. The CBC partly discharged this function by providing facilities for the departments of education to enable them to present school programmes that would supplement and enrich their provincial courses of study. However, from the outset it appeared that school broadcasting could play a wider and equally vital role. Through the CBC's nation-wide facilities, it could contribute effectively to strengthen Canadian unity and to increase the awareness of young people of the importance of their common Canadian citizenship. This idea received considerable impetus from the growth of Canadian nationalism during World War II.

THE CORBETT REPORT

In May 1938 Major Gladstone Murray, the first General Manager of the CBC, commissioned Dr. E. A. Corbett, Director of the Canadian Association for Adult Education, to make a nation-wide report on school broadcasting in Canada. This report, besides surveying developments in Nova Scotia, British Columbia, and other provinces, discussed the question whether at that time "a national direction of school programmes" was desirable. Dr. Corbett thought it would be unwise for two reasons: first, it might arouse suspicions that the CBC intended to "take over" school broadcasting; and second, it would encourage those provincial authorities who were not interested in school broadcasting to regard the whole matter as a recognized federal responsibility and therefore to evade their responsibilities in the matter. "There could be named three Provinces," he said (February 1939) "in which educational officials would welcome any legitimate excuse to side-step the pressure already being brought to bear upon them in this matter by teachers in rural areas."

Before attempting anything in the way of national or regional broadcasts to schools, suggested Dr. Corbett, the CBC should appoint two regional officers, one in the Maritimes, the other in the Prairies, to visit schools and help the education authorities prepare and produce programmes. Also, the Corporation should set up in Toronto a "script exchange," where scripts from Canada, Great Britain, and the U.S.A. could be collected, evaluated, and held for distribution to educational and non-commercial groups. A similar pool of recorded and transcribed material could also be made.

Dr. Corbett's survey was chiefly important as being the first attempt to assemble the record of what had been done up to 1939 in the field of school broadcasting in Canada and other countries. However, his recommendations were necessarily cautious; had he been reporting two years later, when World War II was under way, he would probably have made more drastic recommendations, in the light of Canada's growing national maturity.

Meanwhile, there were other ways in which radio could contribute to the growing national self-consciousness of Canada in education. The CBC lost no opportunities of giving full publicity in its general programmes to educational progress and achievement. School choirs were heard on the air, teachers and administrators described the work done in their schools, school trustees and parents were encouraged to discuss educational issues. In this way the CBC built up for itself the reputation of being interested in, and concerned with, all progressive ideas in the educational field.

CBS SCHOOL OF THE AIR

Early in 1940 a stimulus to school broadcasting in Canada came from outside the country. During the previous decade the Columbia Broadcasting System had been serving the schools of the U.S.A. through its "School of the Air," undoubtedly one of the largest and most imaginative educational projects yet undertaken by radio. Then, under the influence of President Roosevelt's "Good Neighbour" policy, CBS decided to transform its School of the Air into a Pan-American enterprise, seeking to serve the schools of the whole Western hemisphere. The aim was to help the nations of North and South America to a better understanding of one another's culture, history, and ideals.

CBS offered to make available, without charge, the daily programmes of the School of the Air, to be carried on networks in countries outside the U.S.A. Among the first countries to respond to the offer was Canada through the CBC, the Canadian Teachers' Federation, and other educational bodies. Later, a score of other countries including Mexico, the Philippines, and some South American states followed suit; and a Pan-American Council was set up by CBS to meet in Washington or New York to supervise the general policy of the "School of the Air of the Americas," as it was now called.

At the inaugural meeting in Washington, the CBC was represented on the Council by the author, whom Major Gladstone Murray had recently appointed Educational Adviser to the CBC. He pointed out to the meeting that Canada desired to participate in the programme not merely passively, but actively. Accordingly CBS, as represented by their Educational Director Dr. Lyman Bryson, agreed that the participating countries should be invited not merely to broadcast the programmes offered by the School of the Air, but to contribute programmes of their own. The policy was received with general approval.

Canada Contributes Programmes to "School of the Air"

In 1940–41 the CBC first broadcast, on its national network, two of the five courses offered by the School of the Air, those dealing with folk music and literature. The next year Canada undertook to contribute a number of programmes, ten in all, to the School of the Air— programmes featuring the music, literature, history, and industries of the Dominion. These programmes, produced in Toronto, went out not merely to schools in Canada, but to the schools of the U.S.A., Mexico,

South America, the Philippines, and Alaska, and helped the children of those countries to visualize the Canadian way of life, in the same way as Canadian children were learning from programmes originating in those other lands. Among the ten contributions from Canada were programmes describing Ontario hard-rock mining, the Commonwealth Air Training Plan, the lobster fisheries of Nova Scotia, and the folk music of Britain and Canada, besides dramatizations of outstanding children's books by Canadian authors. In subsequent years, Canada contributed annually six programmes to the School of the Air.

Canada's agreement to participate in this programme impelled the CBC to take two further steps. First, a national committee of educators was set up to direct Canada's share in the School.[1] Second, machinery inside the CBC was created to handle the work of organizing the production and promotion of the broadcasts. Mr. C. R. Delafield, CBC Supervisor of Institutional Broadcasts, was asked to take charge of the executive work involved, and the author undertook much of the promotional and public relations work. Copies of the teachers' manual (125 pages) of the School of the Air were widely distributed throughout Canada, and evidence was received that the broadcasts were widely listened to, especially in Ontario and English-speaking Quebec.

During July 1941 Mr. Sterling Fisher, the then Director of the School of the Air, visited Canada and gave demonstrations of the School's programmes in Toronto, Montreal, and Winnipeg. In 1942 the second International Conference of the School was held, by invitation of the CBC, in Montreal. These activities not only aroused keen discussion among teachers and educators about the value of school broadcasting in general, but also produced criticisms of the American School and demands that Canada should increase her own national activity in this field. As we have noted previously, the main criticisms of the American School of the Air were that its programmes were not closely enough related to Canadian educational requirements and that they reflected, especially in the field of social studies, the purely American point of view and tended to overlook the experience and opinion of

[1]The following were members: Miss Lillian Smith, the late Head of Boys and Girls House, Toronto Public Libraries; Dr. C. E. Phillips, editor *The School* (Ontario College of Education); the author, then Educational Adviser to the CBC (Chairman); Miss Jean Browne, Canadian Junior Red Cross; Dr. E. A. Corbett, Canadian Association for Adult Education; Mr. C. R. Delafield, CBC Supervisor of Institutional Broadcasts; Dr. Cliff Lewis, Past President, Canadian Teachers' Federation; Mr. R. W. McBurney, Canadian Institute of International Affairs; Dr. Joseph McCulley, Dominion-Provincial Youth Training Plan; and Mrs. W. G. Noble, Ontario Federation of Home and School Associations.

other countries. For example, in a dramatization of a book dealing with Quebec, travellers from the United States were represented as frequently encountering begging peasant children. In episodes dealing with British history, excessive emphasis was laid upon the harsh behaviour of British redcoats. Misinterpretations of British institutions, and of Canadian attitudes, were fairly common.

Sometimes, on the other hand, the programmes contributed by Canada to the School evoked letters which showed how profoundly ignorant many American listeners were about Canadian conditions. One listener in Pennsylvania, after hearing a broadcast about wheat farming on the Prairies, wrote to express her surprise at learning that wheat could be grown as far north as Manitoba and Saskatchewan!

CANADIAN CITIZENSHIP SERIES

Early in World War II increasing Canadian national consciousness led to the setting up of the Canadian Council of Education for Citizenship, which was supported by public grants. This Council proposed to the CBC the presentation, on the national network, of a series of broadcasts dramatizing scenes from the lives of great Canadians who had contributed to the growth of Canada's nationhood and domestic institutions. Accordingly a series was broadcast weekly from January 14 to February 18, 1942, dealing with six national figures—Joseph Howe, Lord Elgin, Lord Durham, William Lyon Mackenzie, Sir John A. Macdonald, and Sir Wilfrid Laurier. The programmes, each of thirty minutes' duration, went on the air at 10:15 A.M. (in school hours) in the East, and 5:30 P.M. (after school hours) in the West. Afterwards, Mr. F. S. Rivers, the Secretary of the Canadian Council, reported that the broadcasts, which were aimed at the junior high school level (grade 8), had been on the whole very successful. Mr. Delafield, who was responsible for the production of the programmes, considered that they showed that twenty minutes, rather than thirty minutes, would be the ideal length of time for such broadcasts.

CONFERENCE ON SCHOOL BROADCASTS

Two months after the end of the Citizenship series, on April 10, 1942, the CBC called together a private conference of those concerned with school broadcasting in Canada. Under the chairmanship of Mr. E. L. Bushnell, CBC's General Supervisor of Programmes, it was attended by 22 persons from Toronto, Ottawa, Montreal, Winnipeg,

Vancouver, Halifax, and Regina. Dr. Lyman Bryson, CBS Director of Education, New York, was also present. Reports on provincial school broadcasts were given by Mr. Caple (British Columbia), Mr. Morley Toombs (Saskatchewan), Mr. Gerald Redmond (Nova Scotia), and Mr. Aurèle Séguin (Quebec). Dr. Bryson then reported on the CBS School of the Air, whose programmes were criticized by Mr. Caple as "being framed more for the benefit of the general public than for children in the class-room." They contained, he thought, too large a proportion of American and not enough Canadian material. In support of this, the Secretary of the Radio Committee of the Ontario Educational Association, Mr. G. H. Dickinson, declared that the present-day need in Canada was for a better understanding of the province of Quebec, rather than for strengthening relationships with Latin America.

His statement led to a general discussion of the need for a national school broadcast in Canada. Mr. Bushnell suggested that five half-hours should be given up to school broadcasts on the CBC network, of which three should be provincial, one national, and one international. Dr. Bryson agreed but the representatives of some provinces demurred at the idea of limiting provincial programmes to only three out of five.

Dr. Ira Dilworth and the author expressed rather differing views as to the scope and number of the proposed national school broadcasts. The author favoured a series of programmes weekly throughout the winter, devoted to subjects of Canadian national importance. On the other hand, Dr. Dilworth considered that national school broadcasts should be occasional only, in order that they might come to the children as something outstanding and unusual. He suggested, for example, that "it would be good to bring to British Columbia children once a year a programme featuring the breaking of the ice on the River St. Lawrence, or vice versa to bring to the children in the East a programme featuring the salmon spawning season in the rivers of British Columbia." However, the general weight of opinion was that most provinces would prefer a regular, rather than an occasional, national series. Mr. Bushnell also pointed out that it was easier for the CBC to operate on a national network basis than to make its educational broadcasts entirely regional.

To sum up, the Conference reached agreement on four basic points: that one weekly national broadcast to schools during the next winter was desirable, preferably on Fridays; that some programmes of the CBS School of the Air had utility in parts of Canada ("Tales from Far and Near" was the most popular series); that it was desirable that

Canada should continue to contribute some programmes representing the Canadian point of view to the CBS School of the Air; and that information about regional and national school broadcasts should be pooled through the issue of a regular bulletin.

The Conference also adopted Mr. Caple's suggestion that a good subject for the proposed national series would be "Men and Women of Canada." It would consist of dramatizations of achievements by individuals, not necessarily celebrated, who had pioneered or overcome obstacles through displaying co-operative responsibility. In choosing these achievements, emphasis should be laid on social rather than individual problems, and they should be related to the present day. The broadcasts should be each of twenty minutes' duration and should, as far as possible, be contributed and originated from different parts of Canada. On the financial side, it was agreed that each province should be asked, if possible, to contribute its share of the cost.

"HEROES OF CANADA"

Encouraged by the Conference, the CBC Management agreed to produce, during the winter of 1942–43, an experimental series of national school broadcasts, under the title "Heroes of Canada," aimed at the elementary and junior high school level (grades 6–9). To make it clear that the CBC was not taking the initiative on its own responsibility, a letter was sent out to the Department of Education of each province, asking for its approval of the project, and inviting it to contribute one or more programmes to the series at its own expense (cost of script, acting, and music talent involved). The same invitation was also sent to the Canadian Teachers' Federation. A favourable reply was received from the Teachers' Federation and from every province except French-speaking Quebec. In its absence the gap was filled by the provision of two programmes by *Radio-Collège*. Each of the other accepting bodies also selected one or more subjects for inclusion in the series.

In its final form, the series "Heroes of Canada" consisted of 16 twenty-minute programmes, preceded by an introductory preview programme. The series, as announced, was stated to be "of an inspirational character, presenting in dramatized form stories of achievement by men and women of Canada who overcame obstacles and contributed to the life and development of their country, by displaying the pioneering spirit and the sense of social responsibility." "Each broadcast," it was added, "will stress the unity of spirit among the peoples

TABLE I

DETAILS OF "HEROES OF CANADA"

Date	Title	Province contributing	Script-writer	Producer	Place of production
Oct. 8	Preview programme		Harold Foster	Sydney Brown	Toronto
Oct. 9	Sarah Maclure—Telegraphist	British Columbia	Christie Harris	John Barnes	Vancouver
Oct. 16	Richard Uniacke—Dreamer of Union	Nova Scotia		Gerald Redmond	Halifax
Oct. 23	Jeanne Mance—Pioneer Nurse of North America	Quebec (Radio-Collège)	Genevieve Barré	Gerald Rowan	Montreal
Nov. 6	With Axe and Flail—A Story of United Empire Loyalists	Ontario	Alan King	Sydney Brown	Toronto
Nov. 13	Samuel Larcombe—The John Bull of the West	Manitoba	Ben Lepkin and J. W. Chafe	Esse Ljungh	Winnipeg
Nov. 20	Sir Guy Carleton (Lord Dorchester)—Soldier and Peacemaker	Quebec (English-speaking)	Harold Foster	Gerald Rowan	Montreal
Dec. 4	Angus McKay—Protector of Wheat	Saskatchewan	Rowena Hawking	Esse Ljungh	Winnipeg
Dec. 11	Sir Brook Watson—The "Dick Whittington" of the Maritimes	New Brunswick	R. S. Lambert	Gerald Rowan	Montreal
Jan. 15	Frank Oliver—Pushing Fighting Westerner	Alberta	Elsie Park Gowan	John Barnes	Vancouver
Jan. 22	John Stewart of Mount Stewart—Champion of Liberty	Prince Edward Island	Harry Foster	Sydney Brown	Toronto
Feb. 5	With Pack and Pick—Story of Northern Ontario	Ontario	Stan Rivers and H. E. Elborne	Sydney Brown	Toronto
Feb. 12	Sir Samuel Cunard—Conqueror of the Atlantic	Nova Scotia	Dr. Martell	Rupert Caplan	Halifax
Feb. 19	David Stewart—Fighter Against Disease	Manitoba		Sydney Brown	Winnipeg
March 5	Augustus Schubert—The Boy Pioneer	British Columbia	Sally Creighton	John Barnes	Vancouver
March 12	Sir Georges Cartier—A Statesman of Confederation	Quebec (Radio-Collège)	Shulamis Yelin	Rupert Caplan	Montreal
March 19	Egerton Ryerson—Pioneer in Education	Canadian Teachers' Federation	Jean Hambleton	Sydney Brown	Toronto

of all parts of Canada, and will suggest to the boys and girls of today a challenge to attack their own problems in the same spirit as the pioneers of old." Table I gives the complete details of the series.

Thanks to the efforts of the CBC Station Relations Department and to the co-operation of many privately owned stations, the programmes were heard over a network of 41 stations across Canada. Owing to time difficulties, the programmes had to be broadcast (by transcription) at different times in different regions, 11:00 to 11:30 A.M. ADT, 10:00 to 10:30 A.M. EDT, 11:30 to 12:00 noon CDT, 3:00 to 3:30 P.M. MDT, 2:00 to 2:30 P.M. PDT.

The full length of the period provided for national school broadcasts was thirty minutes, but the programmes in "Heroes of Canada" were deliberately limited to twenty minutes each. There remained therefore ten minutes of each programme to be filled. At this time the CBC Central Newsroom in Toronto presented, as an experiment, a short review of the week's news, prepared for students in grades 6 to 9. The exact length of this review varied from five to eight minutes, the time between the end of the review and the beginning of the dramatization being taken up with a short musical interlude.

In addition, on the last Friday of each month the CBC presented a twenty-minute programme to schools depicting outstanding and interesting scenes of life in our country. The programmes entitled "Canadian Horizons" were dramatic in form and were illustrated, where possible, with "actuality" recordings made by the CBC Special Events Department. The four subjects chosen were: "Birth of a Dominion," a play celebrating the Seventy-Fifth Anniversary of Canadian Confederation (Oct. 30); "Douglas Fir, Lord of the Forest," the rain forests of Canada's Pacific Coast (Nov. 27); "Together We Stand," Nova Scotia fishermen's co-operatives (Jan. 29); and "Spring Salmon, the King of Fish," life cycle of the salmon in British Columbia (Feb. 26).

To accompany the broadcasts, the CBC published an illustrated manual entitled *Young Canada Listens* (36 pages), twenty thousand copies of which were distributed, largely through the departments of education to teachers and educators in all parts of Canada.

Some idea of the size of the school audience at this time can be estimated from the following census of listening schools taken by a number of departments of education: Ontario—680 (422 urban, 248 rural); Saskatchewan—300; British Columbia—500; Nova Scotia—150; Manitoba—140; and Quebec (Protestant)—19. In addition, there was evidence of widespread home listening among parents and other adults. The Canadian Home and School Federation publicized the programmes, and discussed school broadcasting at its national, provincial, and local conferences.

Criticisms and Appreciations

The reports received by the CBC from teachers showed that the class-room audience was mainly in grades 6, 7, and 8, though in small rural schools pupils of all grades participated. The best-liked programme in the series, as we have said, was "Sarah Maclure." Of the six that were generally preferred, three came from British Columbia and one each from Ontario, Nova Scotia, and the Canadian Teachers' Federation. The programmes dealing with political, rural, and French subjects were least popular. Of the other national school broadcasts, the news commentary was the most generally appreciated feature. It was found useful for the study of current events, and some teachers asked for it to be continued throughout the year.

In general, the broadcasts were praised by teachers for improving the pupils' English vocabulary and facility for self-expression, teaching children that not all "heroes" are on the battlefield, providing a basis for better listening at home, leading to further research work in class, building up a national spirit, and cultivating children's sense of adventure.

Criticisms were varied. Many teachers, misunderstanding the "inspirational" purpose of the broadcasts, complained that they were not closely enough related to the school curriculum. Others voiced dislike of political and abstract ideas in the broadcasts. Considerable objection was taken to the use of accent and dialect, and to too rapid speaking by actors; also to the excessive use of background music and sound effects in production. It appeared evident that, as school broadcasting developed, there would be needed a special style of scriptwriting and a special kind of production suited to school children.

There was considerable demand for earlier and fuller information about the programmes and their content which was met in subsequent years by expanding *Young Canada Listens*, publishing it earlier in the year, and increasing its distribution. There was also a demand for more advice to teachers on how to utilize the broadcasts in class. The need for this was illustrated by an episode that took place in a Toronto class-room, where an inspector found a teacher teaching his pupils English grammar while they were listening to the Ontario music appreciation broadcasts, "Music for Young Folk!"

A number of teachers referred, in their reports, to difficulties of reception. Investigation showed that in many schools receivers had been installed which were too small to give adequate signal-strength in a class-room. Such receivers, if turned up, created distortion of sound—a deficiency obvious and painful when students were listening to music.

Many of the criticisms revealed by the reports on the first year of national school broadcasts were found to be of permanent significance and required several years of careful and intensive supervision before they could largely be eliminated.

AMERICAN PROGRAMME EXCHANGES

During 1942–43 the CBC continued its collaboration with CBS School of the Air by broadcasting two of its five courses on the national network. With the advice of its national committee of educators, the CBC also contributed to the School six half-hour programmes about Canada. Four of these were dramatizations of well-known Canadian children's books, one was a programme of Canadian music, and the sixth, a scientific programme, was entitled "Seeing the Unseen —the Electron Microscope."

The CBC also co-operated with the newly founded NBC "Inter-American University of the Air," and contributed to it one programme on Canadian history and five on Canadian geography. Several eminent Canadians contributed short addresses to the geography series, including Hon. Angus Macdonald (Minister of Defence for Naval Service), Mr. Charles Camsell (Deputy Minister of Mines and Resources), Dr. Norman Mackenzie (Chairman, Wartime Information Board), Senator Cairine Wilson, and Dr. J. S. Thomson (General Manager, CBC).

On Monday, February 22, 1943, the CBC rebroadcast to schools a transcription, lent by NBC, of the programme dramatizing the Nazi cruelties at Lidice, Czechoslovakia, as written by Edna St. Vincent Millay and acted by a cast of well-known actors. Unfortunately, the promising beginning of programme exchanges with the U.S.A. did not take permanent root. CBS American School of the Air was discontinued in 1948, and NBC Inter-American University of the Air ceased about the same time. When the chief American networks decided to change their policy, and leave school broadcasting to regional and local development, a considerable change in the quality of available school broadcasts in the United States occurred. The place of the networks was taken by a number of local educational radio stations owned by municipal boards of education, universities and colleges, and so forth. Most of these were staffed by "amateurs," and were unable to employ professional script-writers, actors, and musicians. Accordingly, the programmes that they produced did not have the interest for countries outside the U.S.A. that those of the American School of the Air had had. The standard of the school broadcasts produced on these stations proved unequal to the standard of those produced in CBC studios. Exchanges on an equal basis between the two countries became unsatisfactory from the Canadian point of view, and gradually almost ceased. Occasionally, however, goodwill programmes were exchanged with WBOE Cleveland (the station of the Cleveland Board of Education), WBEZ Chicago, and a few other stations.

THE SECOND NATIONAL SCHOOL BROADCASTING CONFERENCE

To evaluate the results of the 1942–43 experiment, a national conference of educational broadcasters from all parts of Canada was called together in Toronto on May 13 and 14, 1943, by Dr. J. S. Thomson (who had succeeded Major Gladstone Murray as General Manager of the CBC). At this conference a letter was read from Dr. H. C. Newland of Alberta, presenting some constructive criticisms and proposals for the future. In his letter Dr. Newland, after paying tribute to the "substantial and valuable assistance that the CBC had already given to the Departments of Education in the Western provinces with their programmes," raised the question of who should control school broadcasting in the future. School broadcasts, he pointed out, undoubtedly fell within the jurisdiction of the provincial departments of education, but so far no arrangements had been made for representing the provinces in this important field. It would, in his opinion, be the proper function of a Dominion board representing the provincial departments of education, rather than the function of the CBC or its appointees, to plan and arrange for school broadcasts that had a national significance and value. The purpose of national school broadcasts was to strengthen Canadian national unity and it was, therefore, important to avoid anything in the nature of "teaching British Imperialism in the English-speaking provinces of Canada, or conniving at French nationalism in Quebec." Dr. Newland recommended the formation of a "Dominion Board of School Broadcasting," nominated or appointed by the departments "to have the responsibility and authority for preparing and arranging for national school broadcasts."

Considerable discussion of his proposal followed, at the conclusion of which the following three resolutions were passed recommending:

(1) That the national school broadcasts be continued and, if possible, expanded.

(2) That copies of resolution (1) be sent to all ministers and deputy ministers of education inviting the co-operation of their departments in the work.

(3) That "the time has come when, with respect to school broadcasting, a more formal arrangement may be necessary, so far as the Departments of Education are concerned. Therefore the Conference asks the CBC to have these Resolutions presented to the Departments and followed up, where possible, by a personal visit."

Dr. W. P. Percival, the Director of Protestant Education in the province of Quebec, suggested that the three resolutions should be referred to the next annual convention of the Canada and Newfoundland Education Association in September 1943.

The CNEA and School Broadcasts

Dr. Percival then drew attention to a newly published *Survey Report on Educational Needs* issued by the Canada and Newfoundland Education Association, quoting the following paragraphs from it:

Full use should be made of the newer tools of education such as the film and the radio.

Broadcasting for classroom instruction must continue to be the responsibility of the provincial departments of education, but there is ample place for programs of an inspirational and broadly informative character which should be regular features of the national network. The Canadian Broadcasting Corporation has made a creditable beginning in this direction. It requires additional personnel and additional resources, however, to make the contribution that will be satisfactory.

That contribution will consist of nation-wide programs and will involve co-operation with provincial authorities in the preparation of specific school lessons. In order that the right type of material may be produced, educational officials should be added to the CBC staff. The most important of these will be a Director of School Broadcasting. He must be an educator of statesmanlike vision who will understand the techniques of broadcasting. He should be able to travel across the Dominion gathering together threads from which Canadian unity will be woven. The Director will associate with himself regional representatives.

Since the preparation of a broadcast requires highly specialized skill and a good deal of expense, broadcast material should be prepared by the provincial departments of education and should be carried on the air through the existing facilities of the Canadian Broadcasting Corporation. The production should be guided by someone who knows the classroom, but the preparation of the script and the presentation should be entrusted to professional broadcasters.

Dr. Percival asked Dr. Thomson to give careful consideration to the CNEA report, and in particular to the recommendation that there should be a Director of School Broadcasting. He hoped that Dr.

Thomson during his term of office would see that this recommendation was put into effect. He himself favoured going still further and creating, within the CBC, a department of education to deal, not exclusively, but particularly with school broadcasting. He also supported the demand for a national committee to co-operate with the CBC in planning school broadcasts. Dr. Percival felt that there had been some discrimination by the CBC in favour of British Columbia, in providing that province with more facilities than those available elsewhere, and urged the CBC to help equally other provinces, especially those that had only modest resources for the development of the work.

Continuation of National School Broadcasts

While these representations in favour of a national advisory committee were being considered by the CBC and the CNEA, the Toronto Conference of May 1943 decided to plan a further series of national school broadcasts for the winter of 1943-44, along the lines suggested by Mr. Caple of British Columbia: (*a*) from October to December 1943, "My Canada" (nine broadcasts giving an imaginative interpretation of the nine provinces of the Dominion to one another); (*b*) from January to February 1944, "The Way of Free Men" (six broadcasts dramatizing the essential principles of the democratic way of life); and (*c*) March and April, 1944, "Proud Procession" (eight broadcasts dramatizing the creative achievements of Canadians in the arts, sciences, construction, community living, etc.). It was also agreed to continue "Current Events" as a ten-minute feature preceding each dramatized programme.

VIII. The National Advisory Council on School Broadcasting

In September 1943 the Canada and Newfoundland Education Association held its annual convention at Quebec, under the presidency of Dr. W. P. Percival. During the convention, Dr. James S. Thomson, on behalf of the CBC, presented a plan for carrying into effect the recommendations passed at the Second National School Broadcasting Conference in May. He proposed the setting up of a council of educators to advise the CBC on its school broadcasting activities, particularly its recently started national school broadcasts. Dr. Thomson was followed by the author who said that, through the proposed council, the CBC hoped to build a real working alliance between broadcasters and educators. After considerable discussion, the CNEA gave its support to Dr. Thomson's plan, and the result was the setting up of the National Advisory Council on School Broadcasting during the fall of 1943.

This Council was created by the CBC, with the endorsement of the CNEA (later the CEA). Once the Council was started, it did not have any direct working responsibility to the CNEA.[1] The constitution of the National Advisory Council, however, was regarded as a joint

[1]In this respect it differed from a similar body later set up to advise the National Film Board on educational films, which was a joint committee of the CEA and NFB, with direct working responsibilities to both.

responsibility to both the CNEA and the CBC. Accordingly, whenever the constitution required amendment, the practice was to refer both to the CBC Board of Governors and to the CEA Board of Directors for their approval.

The establishment of the Council was a novel way of approaching a difficult problem. It provided machinery for co-operation between a federal agency (the CBC) and the provincial departments of education which alone (by the British North America Act) have jurisdiction in education. From the outset, the CBC had wisely decided that it would provide school broadcasts only in partnership with the constitutionally accredited education authorities. To make this partnership effective, the Corporation, again following the wishes of the Toronto Conference in May 1943, set up its own School Broadcasts Department, under a National Supervisor, to co-operate directly with the departments of education, universities, local school boards, teachers, and parents in the development of the school broadcasting work as a whole. The first CBC Supervisor of School Broadcasts was the author, who served in that capacity from 1943 to 1960. Mr. O. C. Wilson was appointed to assist him.

In the first instance, the Council's term of office was set as two years. It was extended by the CBC Board of Governors for further periods in 1946, 1947, 1949, 1951, and so forth, until, in practice, the Council became a "standing" advisory committee to the Corporation.

The Constitution of the Council begins with a statement of basic principles. These are, first, that the CBC is responsible for all that goes on the air, and secondly, that the education authorities are responsible for the utilization in the class-room of what goes on the air. "Therefore the CBC, in presenting broadcasts to schools, wishes to make sure that their educational content meets the approval of the education authorities." Thus the CBC policy in school broadcasting is formulated as being to assist departments of education wishing to provide educational broadcasts to schools on a provincial or regional basis, and to supplement such provincial or regional schemes of school broadcasting by providing, on the national network, school broadcasts designed to strengthen national unity and increase Canadian consciousness among students; also school broadcasts dealing with subjects that are of common interest to the schools of all provinces.

DUTIES OF THE COUNCIL

The functions and duties of the Council are comprehensive, and fall into seven categories: to advise the CBC on the planning of programmes on the national network intended for reception by schools during normal hours; to advise the CBC on programmes relating to educational publicity, Education Week, for example; to advise the CBC on the planning of school programmes to be exchanged with United States and other networks abroad; to advise and co-operate with the CBC on suitable publicity for school and other educational broadcasts; to co-operate with the CBC on matters affecting reception of school broadcasts (i.e., advice to teachers, provision of receivers, distribution of literature, and so on); to collect reports on provincial, regional, and national school broadcasts and to discuss these reports with the CBC; and, finally, to advise provincial governments on changes and new developments in connection with educational broadcasting and to co-operate with the CBC in initiating new experiments in educational broadcasting.

The Constitution states that the Council normally meets once every year, in the spring. At other times special meetings can be called by the CBC in consultation with, or at the request of, the chairman of the Council. Further, each year the Council appoints an executive committee with power to act for it between meetings.

At the outset, the travelling expenses of the members of the Council were met by the bodies appointing them. However, from 1954 onwards the CBC, at the request of the Council, agreed to pay the transportation costs of Council members attending the annual meeting. Travelling expenses of the members attending executive committee meetings were borne in full by the CBC.

The membership of the Council is composed of one representative nominated by each department of education except Quebec, which nominates two (one French-speaking and one English-speaking); two representatives nominated by the Canadian Teachers' Federation; two representatives nominated by the National Conference of Canadian Universities; one representative nominated by the Canadian School Trustees' Association and one by the school trustees of French-speaking Quebec; two representatives nominated by the Canadian Home and School and Parent–Teacher Federation; and one representative nominated by the Canadian Education Association. In addition, the CBC nominates a distinguished educator to act as chairman of the Council. The first chairman so appointed was Dr. R. C. Wallace, then Principal and Vice-Chancellor of Queen's University, Kingston. The CBC also appointed the author, as Supervisor of School Broadcasts, to serve in the capacity of secretary to the Council.

DR. R. C. WALLACE, FIRST CHAIRMAN

Dr. Wallace, one of Canada's foremost scientists, possessed a wise judgment and a broad humane outlook. He was just the man to steer the newborn Council through its difficult initial period. Inevitably a body of this kind, bringing together representatives of nine (later ten) widely separated official educational authorities, a federal broadcasting agency, and a number of unofficial national educational associations, was bound to encounter storms. There were jealousies between East and West, between national and local jurisdictions, between the professional skills of broadcasting and education, and between the interests of official and unofficial educators—all of which required for their conciliation prudent leadership and skilful chairmanship, two qualities in which Dr. Wallace excelled.

The chairman's term of office was nominally for three years. Dr. Wallace served until 1946, when he was succeeded by Dr. W. P. Percival, Director of Protestant Education in the Province of Quebec. He, in his turn, was succeeded in March 1952 by Dr. R. O. Macfarlane, Deputy Minister of Education for Manitoba. In 1954 Dr. H. P. Moffatt, Deputy Minister of Education for Nova Scotia, became chairman, and served until 1958, to be followed in that year by Dr. W. H. Swift, Deputy Minister of Education for Alberta. Thus the Council, after its initial period, had the advantage of being chaired by a leading educational administrator of the highest rank. No deputy chairman was appointed[2] in order to avoid the possibility of a rotation system of chairmanship being established.

MEMBERS OF THE COUNCIL

During the first decade of its existence the Council included among its members many prominent Canadian educators, including the following:

1944–46 Dr. C. F. Cannon, Assistant Superintendent of Elementary Education, subsequently Chief Director of Education for the Ontario Department of Education.

1944–46 Dr. W. J. Dunlop, Director of Extension, University of Toronto, subsequently Minister for Education, Ontario.

1944–52 Dr. W. P. Percival, Director of Protestant Education, Province of Quebec.

1944–55 Dr. B. O. Filteau, French Secretary and Deputy Minister of Education, Quebec Department of Education.

[2]From 1943 to 1946 Dr. W. P. Percival served in that capacity; the appointment was then discontinued.

1944–59 Dr. Lloyd W. Shaw, Director of Education, Prince Edward Island.
1944–61 Abbé Maheux, Laval University, Province of Quebec.
1945–47 Dr. Fletcher Peacock, Director and Chief Superintendent of Education, New Brunswick.
1945–54 Mr. Morrison L. Watts, Director of Curriculum, Alberta Department of Education.
1949–51 Dr. G. A. Frecker, Secretary for Education, Newfoundland Department of Education, subsequently Deputy Minister and Minister for Education, Newfoundland.
1949–59 Mr. George Croskery, Secretary Treasurer, Canadian Teachers' Federation, Ottawa.

On behalf of the CBC also, many of its senior officials attended the meetings of the Council, and took part in the discussions, including the following:

1944–45 Dr. Augustin Frigon, Third General Manager of the CBC.
1944–52 Mr. E. L. Bushnell, CBC Programme Director, subsequently Assistant General Manager and Vice-President.
1948–58 Mr. A. D. Dunton, Chairman, CBC Board of Governors.
1954–   Mr. J. A. Ouimet, Fifth General Manager and subsequently President of the CBC.
1944, 1949–58 Mr. Charles Jennings, CBC Director of Programmes.

INAUGURAL MEETING IN 1944

The inaugural meeting of the Council was held in Toronto on March 9 and 10, 1944, under the chairmanship of Dr. R. C. Wallace. Representatives from the departments of education were as follows: from British Columbia, Mr. Kenneth Caple; from Alberta, Dr. H. C. Newland; from Saskatchewan, Mr. Morley P. Toombs; from Manitoba, Miss Gertrude McCance; from Ontario, Mr. C. F. Cannon; from Quebec, Mr. B. O. Filteau (Catholic) and Dr. W. P. Percival (Protestant); from New Brunswick, Mr. Rolf K. Nevers; from Nova Scotia, Mr. Gerald Redmond; and from Prince Edward Island, Mr. Lloyd Shaw. Representing the Conference of Canadian Universities were Dr. W. J. Dunlop and Abbé Maheux; the Canadian Teachers' Federation, Professor E. L. Daniher and Mr. F. H. Mitchell; the Canadian Federation of Home and School, Mr. L. A. de Wolfe and Mrs. W. G. Noble; the Canadian Trustees' Association, Mr. M. A. Campbell; and the Canadian Broadcasting Corporation,[3] Dr. Augustin Frigon, Mr. E. L. Bushnell, Mrs. Charles Jennings, the author, Mr. O. C. Wilson, Mr. Wells Ritchie, Mr. Aurèle Séguin, Mr. D. B. Lusty, and Mr. Dan Cameron.

[3]CBC personnel present at meetings of the National Advisory Council do not participate in the voting.

Present also was a distinguished visitor from the United States, Dr. Joseph Maddy, Professor of Music Instruction at the University of Michigan, who gave an address to the Council.

Dr. Frigon, in an address of welcome to the Council, said he was convinced that radio must have a place in the school-room, and could play an important part in every child's life. There were two ways of using radio in the school: one, the English method as employed in British Columbia, where radio was used to enrich the school curriculum and where the radio authority worked in close co-operation with the school authorities; the other, that used in Quebec, where the broadcasting authority provided educational programmes on its own responsibility, which were found useful by senior students.

PROGRAMME COMMITTEE SET UP

Three important issues were raised at the Council meeting. First, a Programme Committee was set up, consisting of representatives to examine suggestions for future national school broadcasts and to draft plans for the approval of the Council as a whole. This procedure seemed to operate satisfactorily for the first few years of the Council's existence. Then it began to be criticized. The committee usually met before the meeting of the full Council, with the result that programme plans were discussed twice—once by the committee, and once by the Council. Before long those members of the Council who were not also members of the committee began to complain that they were being treated as mere "rubber stamps" for approving plans previously prepared by the committee. This caused a rift between the representatives of the departments of education, on the one hand, and the representatives of the national, unofficial education associations, on the other.

The question came to a head at the fifth annual meeting of the Council in 1948. At that time Mr. Bruce Adams, one of the two representatives of the Canadian Teachers' Federation, pointed out that duplicate discussions of programme matters were taking place on the committee and on the Council, with consequent waste of time to all concerned. He therefore moved a resolution abolishing the Programme Committee, and making consideration of future programmes the first business of the Council itself. "It is also affirmed," concluded his resolution, "that all members of the Council have the right to vote on all questions relative to the selection of programmes and the compilation of the suggested schedule." The resolution was hotly debated. Representatives of the departments of education expressed the fear that they would be outvoted over programme matters, and that the Council's personnel might be enlarged at some future date, to put departmental representation in a permanent minority. Mr. Dunton denied that the CBC would ever let this happen and promised that, on matters of educational policy and planning, no step would be taken which ran contrary to the expressed wish of the majority of the departments. He pointed out, also, that it would be contrary to the Council's constitution for a majority of its members to deprive a minority of the right to vote. Mr. Adams' resolution was eventually passed by seven votes to six. The rift revealed by the voting gradually closed over the years that followed, as the Council became accustomed to regarding itself as a working unit.

SCOPE OF COUNCIL LIMITED

A second issue of importance was raised at the first Council meeting by Mr. de Wolfe, one of the Home and School representatives, who wanted the Council to take cognizance, not merely of broadcasts to schools, but of all programmes affecting the educational development of children, including children's entertainment programmes. A committee to investigate this matter was set up; but at the second Council meeting, in 1945, Dr. Frigon stated that the CBC could not agree to enlarge the scope of the Council to cover out-of-school broadcasts to children, because it felt that the Council had enough important work to do developing the techniques and improving the quality of school broadcasts. The Council then decided to drop the matter.

FRENCH LANGUAGE BROADCASTS

Another matter that received attention at the first Council meeting was the use of the French language in broadcasting, both for educational purposes and for the promotion of better understanding between French- and English-speaking Canadians. The Council appointed a special committee to inquire into this subject which later prepared a report and circulated it to the provincial departments of education. No positive results followed, however, although the matter was frequently referred to in later Council meetings. Ultimately, with the coming of television after 1954, revived enthusiasm found expression in "Chez Hélène" and similar types of educational television programmes.

EVALUATION OF SCHOOL BROADCASTS

One of the Council's main concerns, from its inception, was to set up a practical method of evaluating the national school broadcasts. The obvious way was to secure reports from teachers who had used the broadcasts in their class-rooms. However, it proved much more difficult than had been expected to secure such reports.

At the outset the CBC, through its School Broadcasts Department and (later) through its Audience Research Bureau, was prepared to attempt evaluations through replies to a questionnaire addressed to teachers and published in the teachers' guide, *Young Canada Listens*. It was also prepared to offer a handsome certificate to all listening schools that agreed to send in such reports on a regular basis. However, the representatives of the departments of education were not willing to agree to any method which depended on establishing direct contact between the CBC and the various schools. It was maintained that this would be, in some way, an infringement on departmental prerogatives. The departmental representatives, while prepared to agree that evaluation report forms should be printed in *Young Canada Listens*, insisted that all returns of such forms should pass through the offices of the individual departments of education before reaching the CBC.

The numbers of such reports which did reach the CBC proved, as the years passed, to be woefully inadequate, rarely exceeding 150 or 200 in any session. Undoubtedly, this was largely due to the natural inertia of teachers towards filling in extra report forms beyond those already required of them. The CBC was debarred by the attitude of the Council from taking any special promotional steps which might have overcome this inertia. On the other hand, the departments themselves, being concerned primarily with evaluating their own local school broadcasts, had little interest in co-ordinating their systems of evaluation with those of other provinces. Consequently, the provincial evaluations, when forwarded to the CBC, could not be co-ordinated on a national basis.

Members of the Council also disagreed over whether the validity of the evaluation depended on the number or on the quality of the teachers who sent in reports. Opinion was about equally divided. After many years' experience, it was decided to conduct evaluations according to both standards. Quantitatively, evaluations were sought through *Young Canada Listens* and through publicity on the air; qualitatively, they were sought through a limited number of selected teachers, covering different types of schools, specially chosen by the departments of education. Neither of these methods proved really satisfactory, however. The problem of evaluating national school broadcasts has never been solved by the Council.

SHORTAGE OF CLASS-ROOM RECEIVERS

Another urgent matter to which attention was drawn at the first Council meeting, and which received the Council's continuing attention, was the shortage of class-room receivers and the poor quality of reception in many schools. A survey made by the departments of education indicated that in 1944 nearly 2,000 new radio receivers were needed in Canadian schools. Many schools were using either borrowed or inadequate receivers.

At the second Council meeting the whole question was examined in detail, with the benefit of advice from representatives of the radio manufacturing industry in Canada and the United States. On the recommendation of the Council's Executive Committee, Professor H. H. Stewart of the Department of Electrical Engineering, Queen's University, Kingston, agreed to represent the interests of education on the newly formed Radio Planning Board, and to act as the Council's technical adviser.

The Council next decided to approach the Minister of Finance, Hon. J. L. Ilsley, asking him to remit the 25 per cent war excise tax on radio receivers, phonographs, etc., in all cases where such instruments were purchased by recognized education authorities. The tax was repealed shortly afterwards.

An attempt was also made to get the War Assets Corporation to grant priority to education authorities in the disposal of surplus war radio equipment. This was done, but much of the surplus apparatus made available was found to be unsuited to school needs.

After the second Annual meeting, discussions were held between the CBC and the Radio Manufacturers' Association regarding the possibility of the production of types of receivers specially suited to school requirements. The manufacturers displayed willingness to provide them, if the school authorities would either guarantee a market or make purchases in bulk. A minority of departments of education were willing to do this, but the majority (including Ontario, where the largest market was) would not and so nothing came of the idea.

At the first meeting of the Council the General Electric Company demonstrated (through a colour film) the possibilities of the new medium of transmission by frequency modulation. At the 1947 Council meeting a further demonstration was given by Mr. F. W. Radcliffe of the Radio Manufacturers' Association of Canada. Subsequently, the Council passed a resolution asking the Ministry of Transport to consider reserving certain FM channels for the exclusive non-commercial use of educational bodies in Canada. At the second Council meeting Dr. Frigon reported that the licensing authority had agreed to do this, and had asked the educational bodies to inform him how many such stations they proposed to operate. Although interest in this subject continued to grow, no school board or department of education actually undertook to apply for an FM licence.

PERFORMERS' UNIONS

From time to time throughout the Council's history, discussion took place regarding difficulties encountered with the American Federation of Musicians in the presentation of school broadcasts. These difficulties covered such matters as the granting of permission for delayed broadcasts in cases where live music talent was employed on school programmes; the granting of permission to make transcriptions of school broadcasts for class-room study purposes; the use of music recordings in dramatized school broadcasts (especially Shakespeare); and the variation between the practices of the "locals" of the union in different parts of Canada. The Council sought the help of the CBC in unravelling these difficulties, which were enhanced by the fact that, at that time, the Corporation had no written contract with the AFM, but operated on a basis of verbal understandings between itself and the union's international (Canada) official, Mr. Walter Murdoch. At the fourth Council meeting Mr. Bushnell was able to announce that the CBC had gained an important concession from the union, allowing for delayed broadcasts under certain conditions. The Council expressed its warm satisfaction with this gain.

Difficulties were also encountered at times with the Canadian Council of Authors and Artists, whose contract with the CBC controlled the conditions under which professional actors, singers, and writers could be employed on school broadcasts. From time to time the CBC, in renewing its contract with CCAA, was able to secure certain concessions on points of detail (mainly connected with relays, repeats, and recordings of broadcasts) affecting school broadcasts as such. However, difficulties remained on such matters as the employment of amateur performers (teachers and students), local variations in fees paid by departments of education to performers, and so on. At the thirteenth Council meeting in 1956, Mr. Neil LeRoy, then President of the CCAA, accepted an invitation to meet the members of the Council, and discussed these problems with them. Although full agreement was not reached, a clearer mutual understanding was achieved, and relations became more cordial thereafter.

TAPE RECORDINGS

As time went on, many schools acquired tape recorders, mainly for the purpose of speech and language training. Some schools, especially high schools whose curriculum was crowded and inelastic, found it convenient to make recordings of school broadcasts "off the air" and subsequently use them for delayed or repeat playing for purposes of class-room study. Permission was sought from the CBC to authorize the making of such recordings, but it could not be given—indeed, the CBC issued several warnings that school authorities making such recordings without permission might become liable to legal action for infringement of authors' or performers' copyright. However, in practice, no such action appears to have been taken, and it gradually became common practice for schools to make such recordings. In some provinces the departments of education or local school boards even undertook to distribute recordings on loan, on a strictly limited and non-commercial basis, however.

From time to time, the need was voiced in the Council for the making of recordings of school broadcasts for distribution on a commercial basis. However, no means could be found of ensuring a sufficient market for the records to justify a record manufacturer in undertaking the risk of production.

COVERAGE OF SCHOOL BROADCASTS

A matter to which the Council gave considerable attention on several occasions (especially in 1950, 1951, and 1952) was the improvement of the coverage of school broadcasts in different parts of Canada, particularly the interior of British Columbia, southern and central Alberta, parts of Saskatchewan and Manitoba, some districts in northern Ontario and Quebec, and some parts of New Brunswick. The difficulties were overcome piecemeal, first by the establishment of high-powered regional CBC transmitters and then by the setting up of "repeater" stations in areas not covered by these transmitters.

A further difficulty arose from the nature of school broadcasting itself. All the programmes were presented during daytime periods, when the signal strength was not as good as at night. Many schools failed to install aerials, and contented themselves with trying to receive the broadcasts on small receivers with a limited range. In some provinces, Saskatchewan, for example, schools relied on receiving their broadcasts from local private stations. But when the high-powered CBC transmitters were built, their coverage often overlapped that of local stations which, in consequence, were apt to discontinue carrying the school broadcasts.

In general, the successful coverage of the whole country could not have been achieved without the generous support and co-operation of the private stations who, with only occasional demurs, freely give time on the air to carry the school broadcasts. These stations were not only those affiliated to CBC's Trans-Canada network but, in several cases, also those affiliated to CBC's Dominion network. By 1960, the total network carrying the national school broadcasts had risen to over 65 stations, the majority of which were privately owned. In some cases, individual stations also promoted the school broadcasts among teachers and, as occasion required, gave studio facilities to make possible the origination of programmes from a local centre to a provincial or regional network.

Upon various occasions, by resolution of the Council, the Chairman of the Council wrote special letters to the privately owned stations expressing the thanks of the departments of education and other bodies represented on the Council for the services rendered by those stations.

In the earlier years of the Council, many distinguished visitors attended its sessions and gave addresses, among them being, in 1945, Mr. Paul Thornton, Education Director, RCA Victor; in 1946, Dr. Franklin Dunham, Chief of Radio, U.S. Office of Education, and Mr. William B. Levenson, Directing Supervisor of Radio, Station WBOE, Cleveland, Ohio (now Superintendent of Cleveland Public Schools); in 1947, Colonel Charles Krug, Canadian Citizenship Branch, Ottawa; in 1948, Mr. Robert B. Hudson, Director of Education, CBS, New

York; and in 1949, Mr. Richmond S. Postgate, Head of School Broadcasts, BBC, London, Mr. Rudi S. Bronner, Director of Youth Education Broadcasts, Australian Broadcasting Commission, Sydney, N.S.W., and Miss Kay Kinane, Federal Producer, ABC, Sydney.

PROGRAMME POLICY

Because the national school broadcasts occupy a thirty-minute period every Friday from October to May, the Council had to decide at the outset how to make the best use of the time. In the opinion of most teachers, twenty minutes represented the optimum time for a school broadcast, enabling the class-room teacher to integrate it properly within a forty-five-minute lesson period. In the case of junior grades fifteen minutes were held to be even more suitable. For senior grades, especially in music and drama, thirty minutes might be acceptable. On this evidence, the Council decided that, normally, twenty out of its thirty minutes should be allocated to broadcasts in fully dramatized form, ten minutes to news or current events broadcasts, and fifteen minutes to programmes of an "actuality" or "interview" type.

Wherever possible, the Council tried to organize its programmes in such a way as to encourage audience (i.e., student) participation in the class-room, partly through the use of "visual aids" accompanying the broadcasts, partly through the stimulation of "follow-up" work in the class-room through the use of *Young Canada Listens* and other publications.

In the light of the definition of the scope and aim of national school broadcasts, the Council came to the conclusion that no useful purpose would be served by trying to present programmes for the first two primary grades: these would be left entirely to provincial school broadcasting. At the other end, it was felt that a limited number of broadcasts for senior high schools should be provided, although it was recognized that the crowded curriculum and time-table in upper school limited the possible size of the student audience likely to benefit from the programmes. National school broadcasts, therefore, have been aimed largely at the intermediate and senior elementary level and/or at the junior high school level.

During the early years, the Council experimented freely with a large variety of topics, covering social studies (geography, history, and current events), art, music, literature (drama and encouragement of good reading), science (history of science, scientific achievements, and conservation) and political and social science (especially the growth of democracy), chiefly with an emphasis on their Canadian aspect. However, no narrowly nationalistic point of view was allowed to intrude itself. In fact, from the beginning, the Council sought to develop, and to incorporate in its schedule, material of an international character, either dealing with the United Nations Organization or developing exchanges with Britain and other Commonwealth countries.[4]

The experiments during this early period were sometimes successful and sometimes unsuccessful. Because every unsuccessful experiment involved disappointing those teachers and students who listened to the broadcasts, it is not surprising that considerable criticism was aroused, which from time to time found expression at Council meetings. Analysis showed that the failures were due in part to unsatisfactory choice of subjects by the Council itself, and in part to lack of experience on the part of CBC producers regarding the specialized character of school broadcasts. On the one hand, topics were sometimes selected which correlated with school curricula, but by their nature did not lend themselves to good radio. On the other hand, producers had not yet got "the feel" of the class-room situation with which they had to deal in their broadcasts. Both parties had to strive to reach a greater understanding, both of the radio medium and of its applicability to class-room requirements.

Provincial Doubts Concerning the National Series

The difficulties were even more pronounced in the case of national school broadcasts than in their provincial counterparts. Some provinces had from the start doubted the value of regular national school broadcasts. In British Columbia particularly there was a feeling of scepticism, inasmuch as in that province the time allocated to the national school broadcasts had to be subtracted from the schedule of provincial school broadcasts. On November 5, 1945, the British Columbia School Radio Committee, meeting in Vancouver, passed a resolution voicing these complaints and asking that they be considered at the 1946 meeting of the National Advisory Council. The resolution ran as follows:

Since the Committee felt that the National School Broadcasts are not satisfactorily fulfilling the purpose for which they were originally planned, it was moved that whereas there is considerable divergence in provincial school curricula,

[4]For further details of such exchanges, see chapter x.

And whereas education is the special responsibility of provincial rather than federal government control,

Therefore be it resolved that the following suggestions be forwarded to the Chairman and the Honorary Secretary of the National Advisory Council on School Broadcasting:

i. National School Broadcasts as at present constituted for in-school listening should be discontinued, except for six or eight broadcasts yearly, of outstanding national importance, to be evenly spaced throughout the school broadcast year. This is felt to be a plan more in keeping with the original conception of National School broadcasts.

ii. Regular programs be provided for out-of-school listening.

This resolution was formally introduced in the Council by Mr. P. J. Kitley and was discussed at length. Mrs. Kern, the corresponding secretary of the Canadian Federation of Home and School, reported that her Federation had, on January 16, 1946, forwarded to the CBC Board of Governors a resolution passed by the Ontario Federation heartily commending the national school broadcasts and expressing the wish that they be continued and developed. The teachers' representatives on the Council, Messrs. Crutchfield and Adams, pointed out that the subjects chosen by the Council in the past had not always been closely enough related to the school curriculum and therefore to class-room needs; also that more instruction ought to be given by departments of education to their teachers on the utilization of school broadcasts. A suggestion by Mr. Kitley that a special advisory committee of teachers from all provinces be appointed to pass all programme plans was turned down on grounds of impracticability and expense. Finally the resolution was defeated on a show of hands, and another resolution, proposed by Mr. Grimmon, was adopted, calling on the Programme Committee of the Council to give the Supervisor greater guidance on programmes.

Reform of Programme Planning

The visit of Mr. Richmond Postgate, in 1949, was particularly profitable to the Council in that it led, through a suggestion made by him, to a marked improvement in the Council's procedure for programme planning. Until 1949, as has been noted, the Council had planned national school broadcasts through its Programme Committee, which made recommendations for approval by the full Council. Because each recommendation had also to be examined by the CBC School Broadcasts Department to assess its feasibility and cost in terms of broadcast presentation, the procedure was found to be cumbersome and often time-wasting. At Mr. Postgate's suggestion, the Council decided to adopt for 1950–51 a procedure found effective by the BBC's Central Council for School Broadcasting, which involved planning ahead for a two-year period. This procedure necessitated the Council, at its annual meeting, first, giving final approval to the programme schedule planned for the coming school year, and, secondly, initiating an outline plan of programmes for the following school year, and commissioning the CBC School Broadcasts Department to work this plan out in detail during the intervening twelve months, for submission to the next Council meeting for final satisfaction. By this procedure, the Council saved itself a great deal of time, while the CBC School Broadcasts Department had the opportunity for mature assessment of the suitability of the programme suggestions made to it by the Council. The "Postgate" procedure proved to be a decided success, and became regular practice from 1951 onwards.

"What's in the News?"

The longest lived and probably the most widely appreciated of the national school broadcasts was the ten-minute series of current events programmes given weekly under the title "What's in the News?" As a rule, there were twenty broadcasts of "What's in the News?" each year. Intended to give teachers and students in intermediate and senior grades background material about an important topic currently featured in the news of the week, the programme was not a "news bulletin" so much as a "news story," with emphasis on the kind of material not likely to be readily available in the daily newspaper. World developments and leading personalities, special items of Canadian, British, and American news (both domestic and international), foreign countries figuring in current happenings, achievements in science and other branches of culture, provided the main features of the programme.

As a general rule, the selection of topics was made in the CBC School Broadcasts Department, in consultation with CBC Central Newsroom, while the preparation of the scripts was undertaken by one of the CBC newsmen, generally as an "extra" over and above his regular work. For the first few years the stories were prepared by Mr. William Hogg, CBC Chief News Editor. On his promotion to National News Supervisor, the task was taken on by Mr. Norman De Poe. Both men made a contribution of the greatest value to the success of the series.

Equally important was the actual delivery of the news broadcasts. It is not enough to have the script written in language and vocabulary which suited the class-room. It must also be spoken in a clear, interesting, and stimulating way, by a voice acceptable to teachers and students. For many years the broadcast was most ably given by one of CBC's senior announcers, Mr. Lamont Tilden.

"What's in the News" is coeval with national school broadcasting itself, having been instituted first in 1942, in association with the pioneer series "Heroes of Canada" and continued on a regular basis from then.

"Voices of the Wild"

Equally successful among national school broadcasts was the series "Voices of the Wild" which began in the fall of 1950 and has been continued ever since. The idea of "Voices of the Wild" was derived from an Ontario series on "Animals and Birds of Canada." The purpose of both series was to familiarize Canadian children with the wild life of their own country, by dramatizing an imaginary situation in which two small children, Betty and Bobby, are taken by their Uncle Jack on field trips to encounter various wild creatures in their natural habitat. Each broadcast centred around one bird or animal, whose call or sounds were identified to enable students to learn them. Some of these sounds were presented on records; others were imitated by the well-known naturalist, Mr. Stuart Thompson. Sounds of other birds were also introduced in the background of each broadcast, and stress was laid on giving information about such aspects of animal life as migration, hibernation, economic importance, and geographical distribution.

"Voices of the Wild" was aimed at the junior grade level (grades 4 to 6) and won immediate and enduring popularity. The first series (1950) dealt with deer, sparrows, geese, beaver, and owls; in successive years the selection was widened to include almost all Canadian wild creatures. Many programmes were repeated two or three times.

The scripts for "Voices of the Wild" have from the beginning been written by Mr. Max Braithwaite, to whom much of the success of the broadcasts has been due.

Occasionally criticism was directed against the broadcasts from academic and official sources. Queries were raised about the precise nomenclature of birds (often found to be a matter of considerable divergence of opinion among experts), about precise conformity with government regulations on conservation and such matters, or about the use of anthropomorphism in presenting animals. The Council, after discussion, decided that it was more important to ensure the popularity and usefulness of the programmes in the elementary class-room than to please the specialists and pundits—which in any case was impossible!

In 1956 and 1957, with the co-operation of the National Museum of Canada, the broadcasts were supplemented by coloured postcards of the animals dealt with, specially prepared for teachers and students, and sold at a nominal price (15 cents per set of six cards). These were found greatly to enhance the educational value of the programmes. In 1959 Canadian Industries Ltd. undertook, as a public service, to make available to class-rooms throughout Canada a free folder containing pictures in colour and supplementary text dealing with that year's "Voices of the Wild." The folder was publicized through *Young Canada Listens,* and over a quarter of a million (257,236) copies were applied for and distributed to teachers for use in their classes. Over seven thousand applications were made by these teachers, of which over one-half came from Ontario, and nearly two-fifths from the three Prairie provinces, the remainder coming from British Columbia and the Maritimes. These figures give an indication of the wide interest taken in the broadcasts by teachers and students in all parts of Canada. Probably no other series except "What's in the News?" has attained such a large listening audience in the schools.

Conservation

"Voices of the Wild" was a science series, with a conservation purpose. A similar programme, also in dramatized form, which proved nearly as effective was "The Adventures of Nanna-Bijou," first broadcast in 1951 for senior elementary grades, and later (1957) adapted and repeated for more junior grades. This series drew on the legendary aspects of Indian life, and treated of the efforts of the Indian demigod, Nanna-Bijou, to follow the instruction given him by the great Manitou when he ordered: "Let there be always harmony and abundant life in the Garden of Canada." The scripts for the first Nanna-Bijou series were written by Mr. Len Petersen, and for the second by Mr. Max Braithwaite.

A valuable clue to the size of the school broadcast audience was afforded by the figures of distribution of a free illustrated booklet on forest management issued by the Abitibi Power and Paper Co. to supplement the broadcasts. Over seventy thousand copies of this booklet (offered through *Young Canada Listens*) were distributed to teachers. The proportionate distribution through the various provinces was similar to that of the "Voices of the Wild" booklet except that a much higher proportion of schools in British Columbia showed interest in the subject of forest management.

Dramatizations of Canadian History

Canadian history, being in its earlier periods largely the story of exploration, lent itself particularly well to radio dramatization. Such dramatizations were able to serve a national purpose by bringing to life outstanding personalities and their achievements and adventures. The first series planned by the Council was the "History of the Hudson's Bay Company" in 1947, for grades 5–9. The broadcasts were supplemented by excellent historical wall maps (in colour) issued free to schools by the Company. The project was repeated in 1960. Thereafter the Council included a series of historical dramatizations for intermediate grades every year. Often the technique, familiarized by CBS, of presenting the broadcasts in the form of eye-witness accounts ("I Was There") was employed, with excellent results. A number of well-known writers, such as Orlo Miller, Tommy Tweed, and others took part in writing the programmes.

History of Art

In three years (1944–45, 1945–46, and 1952–53) considerable success attended series of broadcasts on the history of Canadian art ("Adventure of Canadian Painting" and "Our Canadian Painters") for grades 5–9. In connection with these series, the National Gallery of Canada distributed to interested teachers, at a nominal charge, sets of coloured postcards of the paintings dealt with in the broadcasts. Up to 100,000 of these cards were distributed each year.

Music Appreciation

In music appreciation the Council made repeated experiments with symphonic and choral music, and with opera, to bring to the schools of Canada outstanding live musical experiences which they could not expect to get in any other way in school hours. These included half-hour concerts by the Toronto, Winnipeg, and Vancouver symphony orchestras, performances of Gluck's opera *Orpheus* (1949), Benjamin Britten's *Let's Make an Opera* (1951 and 1961), and Gilbert and Sullivan's *Pirates of Penzance* (1953) and *HMS Pinafore* (1956), and dramatized biographies of great musicians (Haydn, Mozart, Handel, and Schubert), incorporating typical works. These experiments met with varying success. In some cases they were too far in advance of the musical level reached by senior elementary students, or were outside the ordinary range of music teaching in public schools. On the other hand, they were rarely listened to in high schools, which might have benefited most from them, because of curriculum and time-table difficulties. The experiments are still being continued in the hope of finding a formula that will meet the varied needs and standards of students in all parts of the country.

Once each year, on the last Friday before Christmas, the Council invites each province in turn to present a half-hour programme of Christmas music performed by a choral group from one or more of its schools (elementary or secondary). Schools throughout Canada have thus been enabled to hear typical performances of students in a different part of the country. The Christmas music broadcasts were given first on December 23, 1949, by a Winnipeg school choir, and have been continued ever since.

English Literature

In literature, the Council's efforts were concentrated in two different directions, to suit the needs of the high school and the elementary school audiences respectively. At the elementary level, it was felt that the best service that national school broadcasts could perform would be to encourage good reading among students by presenting to them outstanding Canadian books and opportunities for encountering notable Canadian authors in person. Accordingly the programmes included over many years dramatized selections from well-known Canadian children's and other books, for example, "Adventure through Books," for grades 6–10, in 1950–51 and 1955–56, and interviews with and assessments of notable Canadian poets and other writers, for example, "Writers" in 1944 and "Poets" in 1949, also for grades 6–10.

At the secondary level the Council found, after exhaustive enquiry, that the common interests of Canada's high schools could best be served by a systematic full-length presentation of a cycle of three Shakespearean plays, *Julius Caesar, Macbeth,* and *Hamlet* in that order, with one play being presented in the spring of each year. Most provinces, it was found, included at least two or sometimes all three of these plays in their high school courses of study. The Shakespearean presentations proved to be exceedingly popular. In many centres across Canada, young people had little opportunity or prospect of seeing a stage performance of one of his plays, though film versions of some of the plays (such as *Hamlet*) were shown at local movie theatres. Also an increasing number of programmes for the adult audience have been given on the American and Canadian networks, both on radio and on television.

The performances given in the national school broadcasts were rather different in many ways. First, to facilitate class-room needs, the play was presented in half-hour instalments (covering one or more acts) extending over several weeks. Second, a brief introduction and commentary were included along with the text of the drama itself, to interpret the meaning and action of the play so as to increase the students' comprehension. Lastly, both the acting and the production were geared to the students' capacities, and special emphasis was laid on speaking Shakespeare's lines with maximum clarity, and in such a way as to bring out the full value of the poetry. On the other hand, while live music was used to enhance the dramatic continuity, neither it nor "sound effects" were allowed in any way to obscure the meaning of the text. The great popularity of these presentations resulted in curriculum adjustments in certain provinces, to make sure that the particular play studied in high schools coincided with the radio presentation that year.

Canadian Citizenship and Geography

On many occasions, the Council attempted programmes designed to expound or promote, historically or analytically, the ideals of Canadian democracy. The first series, "The Way of Free Men," was provided for grades 6–10 in 1944, while World War II was still continuing. The same subject was dealt with again in 1954 under the title "The Gift of Freedom." Both attempts were only partially successful. The historical growth of democracy certainly lent itself to radio dramatization, but other aspects of the subject tended to be either too political or too theoretical to interest the students at that age level, or, if treated in terms of current actuality, tended to be provincial rather than national in scope.

In general, the Council found it more profitable to teach democracy indirectly, by promoting youngsters' enthusiasm for the practical fruits of democracy as exemplified in Canada's major achievements. These included many products of industrial growth, the extension of enterprise into new fields (uranium mining, opening of the St. Lawrence Seaway) and fresh geographic areas (Labrador, the Canadian Arctic, and so forth), and the cultural achievements of individual outstanding Canadians.

On several occasions (1949, entry of Newfoundand into Confederation; 1954, bicentennials of Alberta and Saskatchewan; 1958, centenary of British Columbia) special commemorative programmes were presented. The Council would have liked to have presented special programmes on the occasion of Royal Visits to Canada. But these were, in all cases, timed to take place during the summer months, after the termination of the school broadcasts schedule.

Over the years, a great number of the national school broadcasts series received awards from the Columbus (Ohio) Institute for Education by Radio, including "Voices of the Wild" (three times), "Our Canadian Bookshelf" (twice), *Julius Caesar* (twice), "Social Studies" (twice), "Children of the Commonwealth," and "I Was There." These awards were made by juries of American educators, who often assessed the programmes in terms of their general educational value, rather than of their specific suitability to Canadian class-rooms. A high proportion of the awards made to the CBC for all types of programmes were given for the school broadcasts, both national and provincial. This fact indicates that the standard of CBC production and performance for school broadcasts was as fully professional as that for other types of programmes.

BRIEFS ON SCHOOL BROADCASTING

In October 1944 the Canada and Newfoundland Education Association, by resolution, asked the Council, in view of the growing importance of school broadcasting, to present a brief to the Parliamentary Radio Committee stressing the necessity of greater support for the provision of provincial and regional, as well as national, school broadcasts.

Later, at the Council's request (for fear of embarrassing the CBC) the CNEA itself agreed to present the proposed brief, which covered 37 pages and made 17 recommendations on such topics as training and employment of teachers in school broadcasts, exchanges of programmes with the United States, improved and protected coverage of school broadcasts on CBC networks, equalization of CBC support for various provincial plans for school broadcasts, development of "supplementary aids," relaxation of union restrictions hampering school broadcasts, acquisition of receiving equipment, development of FM broadcasting, and so forth. The brief estimated that provincial departments of education were spending in 1944–45 about $26,000 on school broadcasts, and that about 20 per cent of all schools in English-speaking Canada were receiving the broadcasts. Four provinces (Nova Scotia, Ontario, Saskatchewan, and Alberta) were assisting schools through grants to instal receiving equipment. The CBC was spending approximately $12,000 on the national school broadcasts, in addition to meeting the costs of wire-line charges for provincial and national school broadcasts, the cost of *Radio-Collège*, and various salary and other overhead costs. The brief was duly presented by the officers of the CNEA to the Parliamentary Radio Committee in the spring of 1946 but, though well received, did not lead to any specific recommendation or action.

On January 10, 1950, in Quebec City, the Council presented through a deputation of its members headed by its chairman, Dr. W. P. Percival, a brief on school broadcasting addressed to the Royal Commission on the National Development of the Arts, Letters and Sciences (Massey Commission). This was one of about a score of briefs presented to the Commission dealing with various aspects of educational broadcasting. The Commission said specifically about school broadcasts:

A number of witnesses offered special comments and suggestions on programmes in which they had a particular interest. School broadcasts elicited enthusiastic praise, as well as helpful criticisms and suggestions. These broadcasts are prepared either by local authorities or in close collaboration with them. Various groups spoke of school broadcasts as helpful to all schools, but particularly to schools in sparsely settled rural areas, where the shortage of teachers often requires inadequately trained persons to carry a heavy burden. More than one brief urged an expansion of school programmes in order to help equalise educational opportunities for the rural and urban child. There were a number of references to voluntary bodies which had helped schools to purchase radios so that they might benefit from the programmes. . . . Certain special representations on school broadcasting came to us from the National Advisory Council on School Broadcasting, and from individuals and local organisations in many parts of the country. Teachers' organisations were very outspoken in their criticisms and suggestions. The Ontario Teachers' Federation thought that the experience of competent teachers should be more fully used, remarking: "Where advice from teachers has been ignored, or scripts written by teachers altered fundamentally, the broadcast has been rendered less valuable for educational purposes." We suspect here a studied understatement. The use of reproduction of paintings in the National Gallery to illustrate talks on the Gallery is appreciated. Teachers suggested that other talks could be suitably illustrated. An adequate transcription service to overcome time-table problems and to preserve valuable programmes was insistently demanded. In general, it was thought that much more money and effort should be devoted to this valuable national service. For 1,000 school broadcasts Canada employs eighteen people and issues nine publications; for a comparable number the British Broadcasting Corporation would employ eighty people and issue fifty-six publications, according to the National Advisory Council on School Broadcasting.[5]

No ascertainable developments took place in school broadcasting as a result of these observations by the Massey Commission.

[5]*Report of the Royal Commission on the National Development of the Arts, Letters and Sciences* (Ottawa, 1951), p. 29.

Again in June 1956 at Halifax, the Council, through its Chairman, Dr. H. P. Moffatt, presented a brief to the Royal Commission on Broadcasting (Fowler Commission) giving a concise description of school broadcasting in Canada and calling special attention to the need for safeguarding and improving the coverage given to school broadcasts by the CBC and the privately owned stations. Specifically, the Council wanted to be assured that, if and when the CBC had only one national radio network, carrying school broadcasts would become a mandatory part of the programme, especially for CBC's affiliated stations. The Commission's Chairman, Mr. Fowler, asked the representatives of the Council who presented the brief several questions, such as why the private stations were expected to carry the school broadcasts free of charge, and why the teachers were not compelled to make use of the broadcasts in their class-rooms. In reply, Dr. Moffat explained that teachers had professional latitude to use the school broadcasts or not use them, in accordance with curriculum needs. The Council's submission to the Fowler Commission was supported by briefs presented by the Canadian Home and School and Parent–Teacher Federation. But the Commission's Report included no direct reference to, or recommendation concerning, school broadcasts.

On the whole, it must be said that the work of the Council, and of school broadcasting generally, made its remarkable growth between 1940 and 1960 with very little recognition and almost no specific help from any parliamentary or governmental body.

CO-OPERATION WITH THE NATIONAL FILM BOARD

Long before school broadcasts began, educational films had found a place as a "teaching aid" in the schools of Canada. However, the material available to the schools was derived mainly from American and British sources. Not until 1951 was the National Film Board of Canada able to set up machinery satisfactory both to itself and to the education authorities of the country for developing specifically Canadian material acceptable to the class-room.

At the 1952 Council meeting attention was called to the desirability of closer correlation between school broadcasts and other audio-visual aids, particularly films and film strips. It was pointed out that, under the aegis of the CEA (CNEA) there had recently been set up an educational advisory committee, along lines similar to the NAC, to advise the National Film Board on matters affecting class-room films. The Council resolved actively to explore the preparation of class-room visual aids that could be used in conjunction with national school broadcasts.

The resolution led to a decision to establish mutual representation on the CNEA-NFB Advisory Committee and the NAC, through their respective secretaries. The author was appointed to represent the NAC (and the CBC) at CNEA-NFB Committee meetings in Montreal and the secretary of the latter committee attended meetings of the NAC. Two projects soon developed out of these visits. The NFB was already engaged on making and distributing to schools film strips on Canadian birds and animals and on Canadian history. Arrangements were made to correlate these with the NAC's two most popular series, "Voices of the Wild" and "I was There." Through *Young Canada Listens,* teachers were given particulars of NFB film strips on topics dealt with in the broadcasts, and encouraged to use them as "follow-up" material.

The second project, first broached in Council resolution in 1954, was for the production of a 16 mm. film on school broadcasting as a joint undertaking by the CBC and NFB, which would deal primarily with the production and effective class-room utilization of school broadcasts. During the next six years this project was the subject of repeated investigation by the educators, the CBC, and the NFB. Unfortunately, the two parties which possessed the resources to make such a film could not be brought to agree on the sharing of the cost involved. The coming of television complicated the situation by enlarging the scope of the proposed film. Although such films, dealing with both radio and film, have been successfully made in Britain, where they have enjoyed wide use as a means of training teachers and enlightening parents on the educational use of the two media, it has not yet been found possible to produce such a film, or films, in Canada. Perhaps the prime, though hidden, cause of the failure is the underlying rivalry that exists between the educators who use the film, on the one hand, and those who use the radio and television, on the other, for class-room purposes. Unreasonable as this rivalry may be, it has served to hold up progress and prevent the two media from aiding each other's development.

CONCLUSIONS

The Council's greatest achievement during the past eighteen years has been to prove that successful co-operation is possible between federal and provincial authorities in the field of education, and that professional educators could work smoothly with professional broadcasters in developing the use of the medium for promoting a keener sense of citizenship and a greater awareness of national accomplishment among the youth of Canada—no mean achievement in a country the size of Canada, and at the present stage of its national growth. However, this important achievement is subject to some qualifications. Over the period, significant changes have taken place in the composition of the Council. At the outset, many senior departmental officers (deputy ministers and directors of curricula) used to attend its sessions. These have now been replaced by more junior departmental officials, with a limited responsibility for taking policy decisions. A similar tendency has been noticeable on the CBC side; senior management has become too busy to attend Council meetings regularly. Consequently the Council has sustained some loss of its original impetus; its meetings have tended towards operational routine.

Secondly, the attitude of reserve which originally characterized the relations of educators towards the CBC has lasted a surprisingly long time, especially in the West. Many instances could be quoted, especially in the promotion and evaluation of national school broadcasts, where this reserve—which might even be termed "suspicion"—has hampered the taking of steps which could have energized and vitalized the popularity and influence of school broadcasts in general, and the national series in particular.

The personnel and funds available for school broadcasting in the provinces have been sadly limited, and the time of the personnel has been fully taken up with the details of provincial school broadcasts, reducing the energy available for planning, promoting, and evaluating the national series, and limiting the concern of departmental personnel for the success of this type of school broadcasting. The personnel has felt that their jobs depended more on the success of their own local broadcasts than upon the success of the national series.

It has often come as an unpleasant surprise to both CBC and other school broadcasters to find what little awareness of the nature and accomplishments of school broadcasting has penetrated to the upper echelon of the Canadian community as a whole, the political leaders, senior civil servants, university presidents, professors of education, and even senior provincial officials. The lack of information has been reflected unfavourably in the budgets made available at various levels, and school broadcasting has not been enabled to progress and take its full place in the educational system.

IX. The School Broadcasts
Department

Shortly before the setting up of the National Advisory Council on School Broadcasting in 1943, Dr. James Thomson, following the advice given by the Canada and Newfoundland Education Association in its *Survey Report on Educational Needs*, appointed a CBC supervisor of educational broadcasts—a title later amended to the more specific "Supervisor of School Broadcasts." He selected for this post the author, who had served for 12 years with the British Broadcasting Corporation and from 1939 onward had acted as the CBC's adviser on educational broadcasts, and who, along with Mr. Charles Delafield, had already played an active part in arousing the interest of educators in school broadcasting throughout Canada.

DUTIES OF SUPERVISOR

The School Broadcasts Department which the author headed formed a part of the Corporation's Programme Division. Its task was to carry out the wishes and plans of the National Advisory Council, to assist the provincial departments of education to put into effect their own local and regional plans for school broadcasts, and to promote school broadcasting generally.

In his capacity as Secretary to the National Advisory Council, the Supervisor had to act as intermediary between the Council and the

CBC management, and to take care that there was full sympathy and understanding on both sides. In his role as Supervisor, he also acted as the Corporation's agent in its relations with the various education authorities individually. He had to negotiate on the Corporation's behalf a series of agreements between the CBC and each individual province (or provinces) desiring to provide school broadcasts, and to make sure that all provinces received equal treatment in their use of CBC facilities.

The Supervisor was regarded as having an over-all responsibility for all school broadcasts produced by the CBC. Specifically his duty was to make sure that no one would criticize the CBC for interfering in matters of educational policy which were the prerogative of the provincial departments of education, that no one would accuse the CBC of favouring one province in the use of CBC facilities or finance, that all school broadcasts were produced in accordance with CBC standards and techniques, and that the costs of school broadcasting would be shared, on an equitable and agreed basis, between the CBC and the education authorities. To carry out these duties, the Supervisor had to enjoy the full confidence of the CBC management and also to keep on friendly terms with the senior officials of the departments of education. Owing to the exceptional conditions in Quebec, however, his relations with the Quebec Department of Education were limited to the Protestant section; his authority did not extend to *Radio-Collège* or any part of the CBC's French network.

RESPONSIBILITY FOR NATIONAL SCHOOL BROADCASTS

As Secretary of the National Advisory Council he had a special responsibility for the organization of the national school broadcasts. Because these were not so directly related to the curriculum as their provincial counterparts, there was all the more need to ensure that they reached a high standard of interest and made an appeal to teachers and students generally. In its planning work, the National Advisory Council concentrated mainly on the selection of topics for various series of programmes and on determining the grade-level of students at which they should be aimed. The Council was not itself in a position to work out in detail the expression of these plans in practical terms of broadcasting technique. At the outset, some Council members distrusted the CBC's ability to do this job in ways that would win educational approval, and they therefore attempted to give detailed instructions as to how it should be done. But the logic of experience gradually convinced the Council as a whole that this task had better be left to the Supervisor and his staff.

As the School Broadcasts Department grew in size, more personnel with class-room teaching experience were employed as programme organizers and producers. They were drawn from all parts of the country and in some cases from the staff of provincial departments of education. Confidence grew, to a point where the Council became content to let the School Broadcasts Department formulate programme plans for its approval and afterwards carry them out along lines justified by experience.

The main task of the Supervisor and his staff was to translate each specific proposal of the Council from terms of pedagogy to terms of broadcasting. As a rule, this involved the use of techniques of dramatization (full or partial) rather than of straight teaching by lecture. The normal unit which worked on each series of national school broadcasts consisted of the programme organizer, the producer, a consultant who was either a research or a teaching expert, and a professional script-writer.

PROGRAMME ORGANIZATION AND PRODUCTION

At the outset, national school broadcasts were produced by producers allocated by the CBC from its general "pool" to specific assignments. This was soon found unsatisfactory, because such producers had been trained primarily for commercial work, or at least for general entertainment work. Before long, therefore, one or more full-time producers had to be assigned solely to the School Broadcasts Department. The first such producer was Miss Kay Stevenson, who was succeeded after several seasons by Miss Lola Thompson. Miss Thompson showed her capacity for working closely under the guidance of the Supervisor and his programme organizers in producing programmes that satisfied the requirements of the class-room. Probably her greatest success was shown in her direction annually, for the benefit of high schools, of certain Shakespeare plays. The plays chosen by the school authorities (*Julius Caesar, Macbeth*, and *Hamlet*) were presented in virtually complete form in four, five, or six half-hour instalments. Many a young Canadian derived his or her first experience of a "live" Shakespeare performance from these broadcasts.

The traditional "independence" of the producer in the studio had to be modified in accordance with special class-room needs, which included clarity of speech, correct grammar and pronunciation, moderate pacing of delivery, no overlapping between voice and music or sound effects, and no overemphasis of background effects for the sake of heightened entertainment value. The producer had to learn that a school broadcast can never be primarily a "show" in the same sense as a broadcast intended for public entertainment. It is astonishing how difficult it proved to build up a tradition of production which would faithfully observe these requirements. It seemed to be necessary to train each new producer afresh.

Within the School Broadcasts Department, specialization of production gradually developed, and in the course of time it became desirable to employ one producer with special musical background, and another with special dramatic background. It would have been impossible to do this but for the fact that the Ontario school broadcasts, as well as the national series, were produced in the Toronto studios. The Ontario Department of Education delegated to the CBC much more responsibility for programme organization and production than did, for example, the Western departments of education. Consequently the increased volume of work falling on the CBC School Broadcasts Department justified the employment of the extra staff referred to above.

Many outstanding musicians, script-writers, and actors took part professionally in the programmes originating from Toronto, including such musicians as Sir Ernest MacMillan, Mr. Boyd Neel, and Mr. Leo Barkin, such writers as Joseph Schull, Tommy Tweed, Len Petersen, and Mavor Moore, and such well-known actors as John Drainie, Bud Knapp, John Colicos, and Barry Morse.

As has been pointed out, the writing of scripts for school broadcasts has been mainly free-lance work, and the supply of competent writers has always been limited. Contrary to natural expectation, practising class-room teachers do not make the best script-writers. Few seem to possess the natural talent for expression in dramatized form or the incentive to acquire the special techniques involved in this kind of radio writing. Professional writers, on the other hand, have proved more successful at meeting the necessary class-room standards. Some teachers who had left the teaching profession and turned to journalism proved extremely successful in combining capacity and skill for the purpose of school script-writing. An outstanding example is Mr. Max Braithwaite, the writer of the most popular and longest lived series of national school broadcasts, "Voices of the Wild."

Producers, writers, and programme organizers have found that one of the best ways to keep school broadcasts in line with class-room needs is to visit schools and observe the actual reception of school broadcasts in the class-room. Many practical hints on improving the techniques of studio presentation are thus picked up.

ENLARGEMENT OF STAFF

At a very early stage, the Supervisor was authorized by the Corporation to appoint an assistant supervisor, who attended particularly to the Ontario school broadcasts, as well as deputizing for the Supervisor during his absence from the office. Mr. O. C. Wilson, the first assistant supervisor of school broadcasts, served from 1944 to 1949, when he was promoted to take charge of the CBC Film Service, in anticipation of the future development of television. Mr. Wilson was succeeded by Mr. T. V. Dobson, a former Ontario public school teacher, who served from 1949 to 1960. Both men were remarkable for their enthusiasm for the work and their loyalty to educational and broadcasting ideals. Mr. Dobson, in particular, gained the confidence of the educators because of his effective work in improving standards of production through constant supervision. He enjoyed the special confidence of the Ontario Department of Education and for several years was popular in Ontario schools with his weekly broadcasts on current events.

By 1960, the staff of the School Broadcasts Department had increased to four executives (supervisor, assistant supervisor, two programme organizers), two producers, and five clerical workers.

EDITORIAL AND PROMOTIONAL WORK

Among the responsibilities of the programme organizer for the national school broadcasts was editing the teacher's manual, *Young Canada Listens*, and any other such supplementary publications. *Young Canada Listens*, published every August by the CBC Publications Branch, is distributed to teachers in all parts of the country, largely through the co-operation of the departments of education. The circulation of the manual grew steadily as school broadcasting gained wider acceptance in the schools until in 1960 it reached 86,500. It also expanded in size to 52 pages, including illustrations. *Young Canada Listens* contained short summaries of all forthcoming national school broadcasts, with suggestions to teachers for follow-up work, lists of recommended books, films and film strips, and other material. At one time, a special abridged edition of the manual was distributed to parents, at the wish of the Canadian Home and School Federation.

This special edition had to be suspended in 1959 for financial reasons.

Besides this editorial work, carried out in the summer months when no school broadcasts were on the air, the senior staff of the School Broadcasts Department often undertook evening engagements to speak at gatherings of parents and teachers, especially in metropolitan Toronto and surrounding parts of Ontario. Similar duties were undertaken by the school broadcasting staff in the various regions of Canada.

Accuracy is more important in school broadcasts than in most other types of radio programme. What comes over the air is invested with authority in the mind of the listener and nowhere more so than in the school class-room. Teachers are quick to spot errors of grammar and pronunciation, as well as errors of fact. And, although school broadcasts are planned as to subject and content by the education authorities, they have always been looked upon by the average listeners—including teachers and students—as "CBC School Broadcasts" and the CBC School Broadcasts Department had generally the responsibility of answering queries on matters of policy and of detail.

The factual accuracy of school broadcasts has not often been called in question, probably because of the care taken by programme organizers, producers, writers, and consultants to verify everything that was said. However, on one occasion, through a stenographic error, the date of the Magna Carta once appeared in *Young Canada Listens* as 1216 instead of 1215, arousing the ire of the Guelph Board of Education which, it transpired, possessed the retentive memory of the proverbial elephant. For many years subsequently, it regularly passed resolutions deploring the historical inaccuracy of CBC School Broadcasts in general—just because of this one incident!

Teachers were as a rule apt to criticize the length and force of school broadcasts, rather than their accuracy. The greatest proportion of their complaints dealt with faults of production and presentation. Parents, who formed a large part of the audience to these programmes, on the other hand made varied and often contradictory criticisms on matters of religious, political, and psychological significance. For example, the broadcasting of fairy stories and traditional songs revealed a strong division of opinion among parents, librarians, and medical men regarding the effects on small children of the "horror" elements in such stories. If they were softened, the complaint was apt to be that the traditional nursery heritage was being violated through bowdlerization. If they were left unchanged, then parents would complain that sensitive children suffered loss of sleep after hearing them. A similar state of affairs existed over racial prejudice, causing the banning of such old favourites as "Little Black Sambo," and the discarding of "Ten Little Nigger Boys." Religious references, too, often inspired strongly conflicting comments from Catholics, Protestants, and Jews.

Additional problems arose out of the desire for authenticity in the presentation of dramatized school broadcasts. Accents and dialects (particularly the British "Oxford accent") were always likely to cause trouble, whether used by speakers or by actors. In the pronunciation of Latin words, also, it was never quite certain whether the "old" or the "new" style was most acceptable to the class-room.

COVERAGE

Another of the department's activities was coverage. The CBC Station Relations Department aimed always to secure, for both the national and the provincial school broadcasts, maximum coverage from privately owned stations, affiliated both to the Trans-Canada network and, where required, to the Dominion network. This coverage was secured on a voluntary rather than compulsory basis. In the majority of cases, the stations readily co-operated, regarding school broadcasts as a public service to their local communities. However, there was always liable to arise some conflict between commercial and non-commercial interests, especially when changes of policy took place in connection with changes in local management.

From time to time a small minority of stations would seek to abandon their responsibility to school broadcasts, often by allegedly making surveys indicating that local schools had lost interest in the programmes. In the early days, when daytime periods were less valuable commercially than evening periods, it was not very difficult to fit school broadcasts into local station schedules. But with the coming of television, daytime radio hours became more precious than evening hours, and pressure grew to "delay" school broadcasts from the most commercially profitable daytime periods. Each proposal of this kind involved detailed discussion and investigation on the part of CBC station relations, CBC School Broadcasts Department, the provincial departments of education, and the schools affected.

EVALUATION

The School Broadcasts Department also devoted much time to an attempt to evaluate the national school broadcasts, along lines approved by the National Advisory Council. To a large extent, its efforts met with frustration.

The main difficulties were due to insistence by the provinces that all national evaluation be done, in the first instance, through the individual departments of education; variations in the statistical methods used by the individual departments, each of which had its own method for evaluating its provincial school broadcasts; and an unwillingness to let the CBC Research Bureau at Ottawa undertake the task of national evaluation.

In spite of the limitations we have discussed, Canadian school broadcasting grew and flourished and in due course began to attract the attention of other countries, particularly those who were beginning to develop the same kind of facilities for themselves. In 1947 the author was seconded by the CBC to serve as radio counsellor to the preparatory commission of UNESCO for a period of six months. Later, after the establishment of UNESCO, a steady stream of visitors came to Canada, mainly from the radio and educational systems of European, Asiatic, and African countries. These visitors wanted to study Canadian school broadcasting at first hand, not only in Toronto but in the other regional centres also. In 1955–56, Mr. P. J. Kitley, Director of School Broadcasts for the British Columbia Department of Education, was seconded to the Ceylon government to advise them on the development of school radio in that country.

PART THREE

Further Developments

X. International Programme Exchanges

Canada is an internationally minded country, partly because of her geographical position and ethnically mixed population, and partly because of her economic and cultural ties with Britain, France, and the United States. Her internationalism is reflected in her school broadcast system which in its initial stages was strongly influenced by elements from the United States (especially, as we have seen, the CBS American School of the Air) and from Britain (BBC school transscriptions).

As long as network school broadcasting lasted in the United States, Canada continued to participate in the Columbia School of the Air by broadcasting over the Trans-Canada network some of its courses and by contributing individual programmes about Canada for broadcast over American networks. But after the demise of the School of the Air in 1948, exchanges with the United States languished. A few programmes of a general character were exchanged with NBC Inter-American University of the Air and subsequently two or three of the American educational radio stations (notably WBOE Cleveland and WBEZ Chicago) exchanged occasional individual programmes with the CBC. However, the expense of telephone lines and other difficulties proved too much for regular action.

From time to time individual educational radio stations in the United States asked for special programmes prepared and transcribed by the CBC School Broadcast Department for their specific use. Also, permission was sought to use transcriptions of other CBC school programmes, both regional and national. However, when attempts were made to establish, on a reciprocal basis, exchanges of programmes and programme material between the CBC and the American educational stations, there appeared a great discrepancy between the rather amateurish productions of many American radio stations and the professional productions of the CBC. When the American broadcasts were offered to provincial educational authorities in Canada they found that their general style of writing and production was too commercial for the taste of Canadian schools. Consequently, the proposed exchanges became too one-sided to be worth developing on a permanent basis. Though attempts were made to arrange with the "Taped Network" at Ann Arbor, Michigan, for the production of school programmes which might be made available on an exchange basis in Canada, this network consisted mainly of adult educational material and nothing suitable for Canadian schools was actually produced.

In general, American radio educators preferred to offer for export to Canada social studies material which reflected the American interpretation of "free enterprise," whereas what was desired by Canadian schools was specific descriptive matter relative to features and conditions of American life.

BBC TRANSCRIPTION SERVICE

During and after World War II, the BBC made available for overseas use copies on transcription of some of her domestic school broadcasts. These were found useful in parts of Canada. From 1945 onwards the supply of programmes was expanded into a regular transcription service. The programmes were of a high quality and often dealt with topics, particularly history and geography, that seemed to tie in well with parts of the curriculum of the Canadian provinces. Accordingly the CBC School Broadcasts Department, from 1945 onwards, agreed to act as a clearing house for the distribution of these transcriptions across Canada. The BBC schedule of transcriptions was offered to individual departments of education for inclusion if they wished in their own provincial series. Because the transcriptions were free, they offered both economical and educational advantages.

Many of the Eastern departments of education, including Nova

Scotia, New Brunswick, Newfoundland, Protestant Quebec, and Ontario, made use from time to time of these BBC school transcriptions; Ontario in particular has regularly scheduled them for a number of years. In Western Canada, however, the position was different. Here complaints were heard that the English accent on the broadcasts made them unintelligible and, therefore, unwelcome to children in Western schools, many of whom came from homes with a different ethnic background. Criticism was directed in both Eastern and Western Canada against the unsuitable length (twenty minutes more or less) of the transcriptions. It was also pointed out that being derived directly from a domestic British source, these transcriptions often contained allusions to facts and persons unfamiliar to Canadian pupils.

At the request of the BBC School Broadcasts Department, from time to time individual programmes of a Canadian character were prepared and produced by the CBC School Broadcasts Department for broadcasting to British schools. The programmes were relatively few in number, however, partly because of a lack of interest on the part of British curriculum authorities and teachers in Canada and its affairs. More is included in the Canadian curriculum about British history and institutions than is included in the British curriculum about Canadian history, institutions, and geography. Every Canadian child learns something about the Magna Carta, Nelson and Trafalgar, the First Reform Act, and the Industrial Revolution. But how many British pupils learn anything about the peaks of Canadian history— the War of 1812, Sir Isaac Brock and the Battle of Queenston, the Rebellion of 1837, and the B.N.A. Act? Similar gaps, of course, exist in their studies of Australian, South African, and Indian history.

INTER-COMMONWEALTH EXCHANGES

By 1945 school broadcasting had developed in most countries of the Commonwealth along the original pattern set by the BBC to a point where programme exchanges through radio were feasible and desirable. The children of Australia, Canada, New Zealand, South Africa, and Britain were growing up with very inadequate information about one another's way of life. Radio, the fastest moving medium of mass communication then know, appeared to offer a short cut towards strengthening inter-Commonwealth ties.

A pioneer series of such exchange programmes entitled "Children of the Commonwealth," aimed at an age level of 11 to 14 years, was produced in 1948, with contributions by the Australian Broadcasting

Corporation. These programmes (of fifteen minutes' duration), which were simple dramatized features of daily life in these countries, were distributed in disc form for delayed broadcasting in Canada, Australia, and Britain. Naturally criticisms and suggestions for improving the programmes were received from the teachers in all three countries. Subsequently, at Canada's instigation, in 1949 the first planning conference of the Commonwealth School Broadcasters was held, in Toronto. It was attended by the heads of school broadcasting from Canada, Great Britain, and Australia.

From then on the project developed steadily, with New Zealand participating in 1949–50 and subsequently South Africa and Ceylon. A fresh series of exchange broadcasts was launched, entitled "Things We Are Proud Of," in which each country presented some aspect (historical or geographical) of its own national life. For example, Britain took the young listeners on an imaginary visit to the Tower of London and Westminster Abbey. New Zealand told of sheep farming and gold mining, Canada of salmon fishing and logging, Australia of its flying doctor service and the rabbit menace, South Africa of lions and diamonds, Ceylon of elephants and tea, and so on. The new series proved encouragingly popular with the schools, but fresh problems soon accumulated, which could be met only in round-table discussions. In 1952 an opportunity for discussion was provided by the Commonwealth Broadcasting Conference in London, when representatives of six countries (Britain, Australia, Canada, New Zealand, Ceylon, and Pakistan) met and thrashed out in detail their common hopes and difficulties about the exchange programmes.

It was agreed that the planning of the programmes should be more closely correlated with the specific curriculum needs of the countries concerned. The productions should be kept close to the grade level of students receiving them. References that might offend local sensibilities (e.g., the subjects of drink or childbirth) should be removed, and phrases whose meaning would vary with the change of locality. This meant careful checking of the script and possibly consultation locally with a native of the country to whom the broadcast was to go. Serious language difficulties were also encountered, for, in spite of the advantage of having a common language for the inter-Commonwealth broadcasts, problems of accent, vowel intonation, vocabulary, pronunciation, and pace of speaking had all to be faced. At the same time, elimination of all differences of speech habit, if feasible, would risk losing one of the big advantages of the broadcasts—the colour and personality of the human voice in its natural aspects.

The educators at the conference concluded that these difficulties could be resolved only by improving the machinery for co-operation and by developing a more flexible type of school broadcast exchange. Instead of each country selecting and producing complete programmes and circulating them on disc to the other Commonwealth countries for inclusion in their own school broadcast schedules, it would be wiser to aim at turning out a programme more closely tailored to the requirements of the individual recipient. For this purpose a "pool" of suggested topics and of material for putting these topics into appropriate form for school broadcasting would be set up. Each participating country would send annually to the "pool" a list of interesting topics suitable for school broadcasting, about which they could offer help with scripts, information, pictures, or recorded material. The collected lists would be circulated to all Commonwealth countries concerned and so brought to the attention of the educational bodies responsible for advising the school broadcasters of these countries. Then if any country decided to include such topics in any of its school series for the ensuing year, it would get in touch with the country making the suggestions and ask for the appropriate material. Later, the BBC Head of School Broadcasts offered to act as co-ordinator and central distributing agent for the exchange of the proposed information and material. For some years the new machinery of exchange was tried and found fairly successful. Gradually, however, new points of difficulty cropped up which could be solved only by further meetings among Commonwealth school broadcast officials. In 1952 it had been recommended that such meetings should be held at regular four-year intervals. However, no further meetings actually took place and the projected inter-Commonwealth exchanges subsequently languished.

However, direct exchanges still continue between neighbouring Commonwealth countries, particularly Canada and Australia. These exchanges, as originally planned by Rudi Bronner (ABC) and the author, were based on a thorough preliminary examination of the curriculum of the two countries to find out what the schools of each taught about the other and how radio could best supplement that teaching. Exchanges were developed of junior grade broadcasts on bird and animal life and senior grade broadcasts on geographical and industrial features of the two countries. The exchanges are still continuing.

Considerable improvement has also been effected latterly in the BBC school transcription service. In 1948 the BBC Overseas Department agreed to produce special "export" versions of BBC domestic school broadcasts for overseas consumption, and to supply these programmes (52 per annum) to Canada and other Commonwealth countries. The broadcasts were to be of a length (fifteen or thirty minutes) more suitable for Canadian network conditions. Allusions likely not to be understood abroad were eliminated, and voices more likely to be acceptable in North America were used in presentation. In 1958 the BBC experimented in producing specimen short programmes of a documentary type dealing with historic sites in Britain. However, these did not meet with enthusiasm when demonstrated at the National Advisory Council meeting. Undoubtedly, over the past fifteen years there has been a slow but steady decline in the desire among Canadian educators for British material, a decline which has been reflected in the slowness of Canadian schools to use the BBC transcription service, in spite of its high quality.

XI. The Canadian Teachers' Federation Inquiry into School Broadcasting

Obviously the attitude of the class-room teacher is vital to the success of school broadcasting. The CBC, departments of education, school boards, and other bodies can plan and present programmes for schools on the air. But because the use of school broadcasts is (rightly) not compulsory, it rests with the individual teacher to decide whether or not his students shall hear the broadcasts, and how to integrate them with his teaching. A school broadcast cannot succeed unless it commends itself educationally to the teacher and is effectively utilized by him. Therefore, the professional opinion of teachers is necessary in planning, presenting, promoting, and evaluating school broadcasts.

In most of the early local experiments with school radio in Canada between 1925 and 1943, teachers made an important contribution, through advice, performance, and evaluation. In British Columbia school broadcasts owed their original impetus to the experiment conducted by the Okanagan Valley Teachers' Association. As departments of education became interested, one by one, in school broadcasting, they usually sought the advice of teacher committees, on which the local provincial teachers' association was represented.

When national school broadcasts were projected, active support was given by the Canadian Teachers' Federation through its Secretary–Treasurer, Dr. C. N. Crutchfield. The Federation accepted the CBC's invitation to sponsor one of the programmes ("Egerton Ryerson, Pioneer in Education") in the inaugural series "Heroes of Canada" in 1942–43. Subsequently, the Federation passed a resolution at its annual conference, calling for the institution of a regular national service of school broadcasts. When the National Advisory Council on School Broadcasting was set up in 1943–44, the Federation was allocated two representatives on the Council; one of these was usually the Federation's secretary, the other a teacher in the Toronto area.

On the Council, the CTF representatives played an increasingly active role. They pressed for the closest possible correlation of the programme plans to curriculum needs, and for the better training of teachers in the utilization of school broadcasts. The CTF representative was largely responsible for the basic decision of the Council in 1948 that planning of national school broadcasts should be done by the Council as a whole rather than by a committee composed of representatives of the department of education.

RESEARCH INTO SCHOOL BROADCASTING

In 1950 the Council (at the suggestion of Mr. Philip Kitley) passed a resolution favouring research into the basic philosophy, aims, and accomplishments of school broadcasts. The Council invited the Canadian Teachers' Federation to undertake this project; and in 1951 the CTF agreed to do so, and to publish the resulting report and make it available to the Council. The directors of the CTF appointed a Radio Research Project Committee, under the chairmanship of Miss Kathleen Collins, whose members were largely teachers in British Columbia. However, at all stages of this project, close liaison was maintained between the Council, the Federation, and the Canadian Education Association.

By January 1952 the Committee defined the scope of the enquiry as covering: educational outcomes (to see how the objectives of school broadcasting were being realized); class-room use (including school organization, size of class-room audience, reception, utilization, and follow-up of broadcasts); and presentation (planning and techniques of presentation of programmes). The Committee drafted questionnaires, 6,500 copies of which were distributed by the Federation's Head Office in Ottawa, through provincial teachers' organizations to teachers in all parts of Canada. The resulting replies were tabulated and analysed by Dr. F. Douglas Ayres, the Research Director of the

CTF, who also drafted the Committee's report. Consultation was maintained with the School Broadcasts Department of the CBC.

THE CTF SURVEY

The Survey of Radio in Canadian Schools (75 pages), issued by the Canadian Teachers' Federation in 1956, covered a good deal of ground. It found that approximately 75 per cent of the schools were equipped with radio, and that most of those still without radio were in rural areas. Even in larger schools, however, there was only one radio receiver available, on the average, for every five teachers. The Committee recommended that the regulations of departments of education be made sufficiently flexible to ensure that wherever there was real need for equipment (radios, tape recorders, and central sound equipment), assistance would be available. Provision should also be made for ensuring that such equipment be of suitable standard.

The Committee recommended that the objectives of school broadcasts should be clearly defined and published in teachers' guide books. It recognized that not all school broadcasts must be correlated in the same way and to the same degree to the school curriculum.

It was found that about 40 per cent of the teachers frequently did little or no planning in preparation for individual programmes and that only 21 per cent of teachers had had training in the use of school broadcasts. The *Survey* recommended that general suggestions for utilization and comprehensive outlines of forthcoming programmes be included in all teachers' guide books, and that teachers ought to "recognize the value of adequate preparation for school broadcasts, be thoroughly familiar with the various techniques that can be used during and following broadcasts, and make the fullest use of relevant material before, during and after the actual broadcast." Responsibility for effective utilization did not fall alone on any one group of educators, but rather on all. The Committee recommended that all teacher education institutions should give instruction in the effective utilization of school broadcasts.

Finding that in the larger schools only 39 per cent of the teachers were using school broadcasts, the Committee recommended that better organization for receiving the programmes be undertaken in each school and that more use be made of tape recorders to overcome the disadvantages of school broadcasts coinciding with class period times.

The school broadcasts were favourably commented on both by the teachers and by the pupils (40 per cent of the student population) who listened regularly to them. They enjoyed also a large and appreciative audience among parents. The Committee therefore strongly recommended that the CBC, the private radio stations, and the departments of education should continue their support of school broadcasts, both national and provincial; and that the National Advisory Council should continue to function and, upon the coming of television, extend its scope to cover the new medium. "The nature of the CBC and its policy of co-operation with the Departments of Education is primarily responsible for the healthy state of school broadcasting in Canada."

Teachers in Ontario and Quebec were found to favour afternoon, rather than morning, periods for school broadcasts, and the CBC was asked to change its scheduling of school broadcasts accordingly.

Greater variety of subjects dealt with on school broadcasts and more interprovincial exchange of programmes were recommended. The length and number of programmes in series should be kept under continuous review.

Approaches should be made to private stations on the Dominion network of the CBC to persuade them to carry school broadcasts, wherever necessary, for good coverage. If such requests were not favourably received, the CBC should invoke its compulsory powers under the Broadcasting Act.

It was found that 20 per cent of the teachers using school broadcasts had no access to a provincial guide book and 33 per cent of teachers of grades 3–9 had no access to a national guide book. "Obviously a large number of teachers are using school broadcasts without adequate direction. When a teacher does not know the content of a broadcast beforehand, the benefit the pupils will obtain from that broadcast is minimized." The Committee, therefore, recommended that departments of education make provision for more thorough distribution of these guide books.

Nearly 1500 teachers made suggestions to the Committee on points of presentation. The Committee picked out from these the following, which it recommended to the attention of producers of school broadcasts: avoidance of "high-pitched female voices" on the air; greater insistence on clarity of speech; restraint and care in the use of dialect, sound effects, and music in dramatized programmes; use of Canadian voices on international programmes presented in Canada.

The Committee recommended further research into school broadcasts to answer the following queries. Is the demand for certain programmes spontaneous or created by providing those programmes? Are certain subjects being allowed to monopolize the limited broadcast time? Should more time be allotted to school broadcasts each day? Are all programmes sufficiently utilized to warrant their continuation? Are longer series more apt to be used than shorter series? What effect does the method of programme presentation have on usage?

The need for more frequent and more thorough evaluations of school broadcasts by teachers was stressed. It was recognized that some of the weaknesses referred to in the *Survey* were due to limitations of the sound medium in presenting ideas and material that were predominantly visual.

Results of the Survey

The *Survey* was well received, particularly by the local branches of the Canadian Teachers' Federation and in departments of education, where it was taken as evidence of the substantial progress achieved to date in school broadcasting. Some of the specific recommendations made in the survey have since been implemented, including better organization for reception in schools and greater use of tape recorders, afternoon periods for broadcasts in Ontario and Quebec, and improvement in techniques of presentation. The problem of adequate reception also received considerable attention from the National Advisory Council which requested the CBC, as soon as a single radio network schedule was adopted (in accordance with the recommendations of the Fowler Commission), to place the school broadcasts in a category of programmes that would be mandatory for affiliated stations to carry.

From 1954 onwards, the attention of educators was increasingly focused on the new medium of television. Here, too, the CTF played an active part. When the National Advisory Council asked the CBC to undertake its initial experiments into school television, a special committee of members of the Council was set up to plan and guide the experiments. At that time no department of education was sufficiently convinced of the merits of school television to wish to take the lead. Accordingly, the chairmanship of the Television Committee of the Council was accepted by Mr. George Croskery, the Secretary-Treasurer of the CTF. He threw himself whole-heartedly into his work and played a leading part until in 1958 illness forced him to relinquish his position. By that time the CTF had appointed its own audio-visual education committee, which concerned itself particularly with television. Member bodies of the Federation, such as the Ontario and other provincial teachers' federations, also established their own television committees, and supported the various projects that took shape in different parts of Canada.

On the whole, the influence of the teachers' organizations has always been strongly in favour of using radio and television to improve the effectiveness of teaching. Though approving of their use for "enrichment," many teachers have remained sceptical of plans for using television for "direct teaching," fearing that it would lead to a mechanization of the teaching process and the down-grading of class-room teachers, part of whose function might be usurped by school telecasts.

XII. The Family Audience

School broadcasts are the only type of regular class-room activity which can be heard and followed in the home. For this reason they have always attracted a sizable adult audience, in addition to the student audience for which they are specially intended. Because of the time of day when the programmes are aired, the adult audience is largely feminine. There are also some children missing school for illness or other reasons, and even a sprinkling of men who happen to be listening on car radios when travelling or en route to or from their work.

The adult audience is important because it tends to set the climate of "public opinion" about school broadcasts and their value.[1] Mothers of families, in general, have shown serious concern with the fare offered to their children over radio and television. Many inquiries and surveys have been conducted to determine whether the mass media exert an uplifting or a degrading influence upon the thinking and behaviour of young people. Their findings, in Canada, have always tended to stress the contrast between instructional broadcasts for children, and those which aim purely at entertainment. By and large, the school

---

[1]On various occasions, outstanding school broadcasts were repeated during the evening, generally as part of the "Wednesday Night" programme on the Trans-Canada network. Chief among these were the annual productions of Shakespearean plays, *Julius Caesar*, *Macbeth*, and *Hamlet*, for the high school audience.

broadcasts have appeared to be the type of programme that parents would wish and encourage their youngsters to hear.

HOME AND SCHOOL SUPPORT FOR NATIONAL SCHOOL BROADCASTING

The Canadian Home and School and Parent-Teacher Federation was one of the first educational bodies in the country to give its interest and support to school broadcasting, both at the national and at the local level. It did so by inviting speakers from the CBC to its annual conventions, holding discussions in its provincial and local associations, and passing resolutions favouring school broadcasting, for forwarding to official educational and political bodies. The Federation and its provincial associations also appointed "conveners" for radio, television, and other audio-visual aids, who helped to promote interest in the work, particularly if the convener happened to be an energetic Home and School leader. However, as each convener held office for only three years, there was apt to be discontinuity in their effectiveness.

When the National Advisory Council on School Broadcasting was set up in 1943, the Canadian Home and School Federation was invited to send two representatives to attend. Usually, these were the president or a vice-president, and the national radio convener for the time being. The Federation's representatives on the Council were less prominent (as might be expected) in programme planning than in forwarding the promotional aspects of school broadcasting, or in urging the need for more school receivers. In the latter field, local Home and School associations were often active and helpful. Some associations pressed local school boards to make the necessary financial provision for sets; others themselves helped raise funds for equipping local schools with one or more receivers.

CO-STUDY EXPERIMENT

Although the Federation left the business of planning programmes mainly to the departments of education and the teachers, it frequently concerned itself with evaluating the results of the broadcasts, especially in so far as they affected students in their own homes. The most interesting of these evaluations was the experiment on "Co-Study" undertaken jointly in January and February 1945 by the Ontario Federation of Home and School and the CBC School Broadcasts Department. "Co-Study" was defined as "a means whereby parents, through radio, can share in their children's education at school." Basically it involved the continuation in the home, between parents and children, of the "follow-up" work started in school after hearing the school broadcast. Because school broadcasts were not "lessons" in the orthodox sense, but intended to stimulate the imagination and enrich the curriculum, any interested parent could take a hand in carrying the process further.

Three schools in the Greater Toronto area—Rosedale and Blythwood public schools in Toronto, and R. G. MacGregor public school in East York—were chosen for the experiment. A series of five national school broadcasts on "The Adventure of Canadian Painting" was made the subject of the experiment. Lists of children who would be hearing the broadcasts in class were drawn up by the school principals, and a voluntary panel of mothers of these children was formed who undertook as far as possible to listen to the broadcasts at home and afterwards to follow them up with their children. A memorandum, *Hints on Co-Study*, was furnished by the CBC and supplied to the parents. The hints were advice on how to inform themselves properly about the subject (Canadian art) and how to follow it up by reading, conversing with their children, visiting art exhibits, making collections of prints, and doing art work in the home. Copies of the colour prints distributed by the National Gallery of Canada at Ottawa in connection with the broadcast (see p. 149) were used in the three schools and by the parents who took part in the Co-Study experiment.

Approximately 70 per cent of the parents who reported on the results of the experiment said that the broadcasts had stimulated discussion in the home and had permanent educational value. About 20 per cent indicated that the broadcasts had been followed up by visits to the Art Gallery. The chief value of Co-Study was that it increased the interests of parents in their children's formal education and encouraged them to take part in it.

Other experiments in Co-Study took place later elsewhere in Canada, especially in the Martitimes, in connection with grade 8 French broadcasts.

HOME LISTENING SURVEY

During the winter of 1954–55 the Canadian Home and School Federation, in collaboration with the CBC, conducted a country-wide home listening survey of school radio, covering the period from October to December 1954. The survey was carried out by the Federation's Audio-Visual Education Committee, under the chairmanship of Mr. Clifford E. Edwards, a school inspector in Nova Scotia, with assistance from Dr. M. V. Marshall, Head of the Educational Department of Acadia University, and three students of the university, Mr. Neil Johnston, Mr. David Nathanson, and Mr. H. E. Comstock.

The purpose of the survey was to determine the extent to which parents listen to school broadcasts, and to get parents' opinions of the programmes and their suggestions for improving them. Over 13,000 questionnaires were sent out to selected Home and School associations, but only 1751 (14 per cent) were returned, less than 1 per cent of the total Federation membership. Over two-thirds of those reported that they had listened to the school broadcasts during the period of the survey. Top priority in popularity among the programmes was given in nine provinces to "Kindergarten of the Air." Sixty per cent indicated that, either frequently or occasionally, they discussed school broadcasts with their children at home. Few of the parents saw the first experimental series of school telecasts presented by the CBC in 1954. Many questionnaires expressed special interest in the CBC programmes arranged for parents, such as "School for Parents" and "Way of a Parent."

The survey also showed that the average number of radios in Canadian homes was slightly under two, and that in over half of the households the radio receiver was in the living room. The most popular time for younger children to listen to radio or watch television was from 4:00 to 7:00 P.M., and for older youngsters from 6:00 to 9:00 P.M.

The survey indicated that many adult listeners were not fully aware of what programmes were being given, even in their local broadcasting schedule. Apparently they did not get either *Young Canada Listens* or their local provincial teachers' manual. It was suggested that the CBC should devise some method of keeping adult listeners interested in school broadcasting better acquainted with the current programmes. As a result, the CBC undertook to publish in 1957 a separate "Home Edition" of the national teachers' manual, *Young Canada Listens*. The popularity of the edition was shown in its circulation of 20,000 copies, mainly through local Home and School associations. The Home Edition was again published by the CBC in 1958 but, as we have said, for reasons of economy its issue had to be suspended in 1959. At the Federation's request, the National Advisory Council on School Broadcasting passed a resolution regretting its discontinuance and pressing for its reissue as soon as conditions permitted.

KINDERGARTEN OF THE AIR

At the fifth Advisory Council meeting in March 1948 the Canadian Home and School Federation presented a resolution (which was passed) strongly supporting an experiment with a "Kindergarten of the Air" which had been launched on a temporary basis a few months previously by the CBC and the Junior League of Toronto. This led to the regular provision of an important daily radio kindergarten series. The idea of "Kindergarten of the Air," derived from Australia where it had proved highly successful, was based upon the belief that radio broadcasts could give to young children in the home a preparatory training for life in school.

In some parts of Canada kindergartens had been established for several years; but in other parts, especially the rural districts, they were conspicuous by their absence. When the CBC School Broadcasts Department broached a plan for an experimental radio kindergarten, the Junior League felt that there was an opportunity to fill a real educational gap. Accordingly the League agreed to make an experiment possible over a CBC network by providing the funds to pay for the scripts and performance of a 13-week series, running three days a week during the winter of 1947–48. The performer, to whose personality much of the success of the series was due, was Dorothy Jane Goulding (Mrs. W. Needles), a Toronto girl, educated in England, Vienna, and Toronto, an Associate of the Royal Conservatory of Music in Toronto, who held a teaching certificate from the Toronto Normal School. Miss Goulding had had experience of kindergarten teaching, and had specialized in the presentation of children's musical and dramatic programmes, both on the stage and over the radio.

The experimental series proved very popular everywhere, especially among educators in the Maritime provinces, and among parents in all parts of the country. The Maritime School Radio Committee and the Canadian Home and School Federation both presented resolutions of support, and called upon the CBC to take over full responsibility for the programme, and extend its coverage across Canada. This Mr. Dunton, the Chairman of the CBC Board of Governors, and Mr. Bushnell, the Director of Programmes, promised to do. Accordingly, on October 4, 1948, "Kindergarten of the Air" was inaugurated as a regular CBC programme and continued on the air until May 1961. Its aim was clearly stated in *Young Canada Listens* for 1948–49: "All over the country there are homes containing children who are not yet ready, or able, to go to school. 'Kindergarten of the Air' is designed to meet the needs of these children by providing them with pre-school training in games, songs and useful activities."

"Kindergarten of the Air" was planned with the advice of kindergarten experts and representatives of the Canadian Home and School Federation, the Junior League, and the Federation of Women's Institutes. Each broadcast lasted fifteen minutes, and the series was heard daily from Mondays to Fridays, usually in a mid-morning period. It included an introductory song theme, exercises (stretching, marching, dancing, etc.), learning songs, listening to a story, and participating in play activities. Parents were asked to co-operate by providing materials such as paper, crayons, and blunt scissors. The children were always given some suggestion for indoor or outdoor activity of a constructive nature—such as collecting specimens of leaves and flowers, or making themselves useful about the home. Many of the songs and stories used in the broadcasts were of a traditional or folk character.

From the outset it was felt that, although in Miss Goulding (known to the children as "Dorothy Jane") the CBC had founded an outstanding performer, nevertheless the onus of a daily programme called for sharing the burden between two persons. The second performer on "Kindergarten of the Air" was Miss Sandra Scott, a well-known professional radio actress. Miss Goulding prepared her own scripts, but those of Miss Scott were written for her by Miss Hazel Baggs, a well-known instructor in kindergarten practice at the Toronto Normal School and a leading member of the Canadian Association for Childhood Education. She was assisted by her friend and colleague, Miss Gladys Dickson. In 1949, Sandra Scott was succeeded by Ruth Johnson (Mrs. Peter Francis), a graduate of the University of Toronto with extensive training as a singer and musician. The combination of "Dorothy Jane" and "Ruth" continued until 1956, when Miss Goulding retired. She was succeeded by "Joy" Maclean (Mrs. Maud Maclean). Miss Baggs and Miss Dickson wrote the scripts for all performances from 1956 to 1961.

Many amusing episodes are recorded in connection with "Kindergarten of the Air." For example, "Dorothy Jane" used to suggest to her young listeners that they try growing cuttings from household vegetables in saucers of water, and advised that they be kept in a cool place away from direct sunlight. An indignant father wrote that, coming home from work at midnight and feeling hungry, he opened the refrigerator and reached inside to get himself a snack, but brought his hand out gripping a large bunch of sprouting carrot-tops! An exasperated mother phoned to "Dorothy Jane" one day after the broadcast, to complain that her child was lying on the floor beside the radio, refusing to get up. In the course of the broadcast, the kindergarten class had been instructed to lie down in front of the radio, and "Dorothy Jane" had signed off without cancelling her instruction. The child refused to move without her kindergarten teacher's order!

Many small controversies and minor disputes in which teachers, doctors, parents, and psychologists were involved, marked the course of "Kindergarten of the Air." "Dorothy Jane" often made up her own stories, many of which were taken from fairy tales and myths. As usual, the purists would rage if there were any departure from the strictly orthodox versions; on the other hand some psychological experts continually questioned their value and appropriateness. Religious and racial issues also provided unexpected "snags." A member of the Lord's Day Observance Society in Saskatchewan once bitterly complained that Dorothy Jane was encouraging infants to violate the sanctity of the Sabbath by advising them to help their parents in the garden on that day. "Do it on a *Sunday* morning," she was alleged to have said, but the complainant was somewhat abashed when shown the script, which ran "Do it on a *sunny* morning." How far religious celebrations should figure in the kindergarten broadcasts was a matter of frequent argument between Jewish and Christian family listeners. Jewish mothers protested against their children being taught "the Christ story" at Christmas and Easter. But when they were left out, "Kindergarten of the Air" found itself publicly condemned for "godlessness," and complaints were made that Christmas was being relegated to present giving and Easter to bunnies. Eventually, an informal compromise was reached whereby the programme's allusions to religion were limited to the Nativity Story and the singing of carols during the week before Christmas.

As time went on, the appeal of "Kindergarten of the Air" widened. It had started as a programme intended purely for the home audience. But gradually its usefulness was discovered by nursery schools, kindergarten classes, and even the occasional grade 1 class in public school. Group listening increased as kindergarten teachers discovered that the broadcasts could be included as a useful and stimulating episode in their class routine every morning.

No statistical estimate could be made of the audience for these broadcasts. However, it is an old broadcasting adage that "you never know how popular a programme is, until someone suggests it may go off the air." In 1957 a report originating in Alberta spread through Western Canada to the effect that the CBC was planning to discontinue "Kindergarten of the Air." There was no truth in the rumour, which was based upon a mistake, but it had one good result—it showed how much the programme was valued in the West. Over one hundred letters came into the CBC School Broadcasts Department protesting the reported termination of the broadcasts. Most of them were in the form of resolutions passed by well-attended meetings of local Home and School associations in Alberta. But there were also many letters from individual parents, of which the following was typical: "The children in rural districts have no other chance for pre-school training but this programme. Even for city children, it has been a real joy. My two have enjoyed it in the morning, even after they were old enough to be able to attend kindergarten in the afternoon. I trust you will see your way clear to continuing this valuable programme."

Although "Kindergarten of the Air" was not in any formal sense a "school broadcast," it came under the supervision of the School Broadcasts Department because of its pre-school training aspect. It figured regularly as an item on the agenda of the National Advisory Council and in 1957 the Council adopted a recommendation made by the Executive Committee of the Canadian Home and School Federation urging the CBC to provide, in addition to "Kindergarten of the Air," a televised educational kindergarten programme, produced in Canada, following the basic principles which had proved so successful with "Kindergarten of the Air." This led naturally to the establishment by the CBC of a visual counterpart to "Kindergarten of the Air" in the form of daily telecasts entitled "Nursery School Time." In 1961 "Kindergarten of the Air" was remodelled and superseded by a new daily radio pre-school programme entitled "Playroom" which was intended not as a kindergarten type of broadcast, but "to serve directly the cause of development in the very young, ranging from age three to age five."

XIII. First Experiments in School Television

Television made its debut in September 1952 when the first CBC television stations were opened in Montreal and Toronto. By June 6, 1961, the CBC was operating 16 stations (plus 4 rebroadcasting stations) of its own, while private enterprise was provinding 48 stations (plus 22 rebroadcasting stations). In June 1959 the longest television network in the world was completed, stretching 4,200 miles from Victoria, British Columbia, to St. John's, Newfoundland. Two CBC television networks, one in the French, the other in the English language, and one independent commercial network are now in operation.

The position of the CBC in television was rather different from what it had been in radio. When the Conservative government, under Mr. John Diefenbaker, took office in 1957 it introduced considerable modifications in the broadcasting system. The Broadcasting Act of 1958 separated the regulatory from the operating functions of the CBC. It transferred the general control over broadcasting and the function of recommending broadcast licences from the CBC to a new body called the Board of Broadcast Governors(BBG), which assumed a role roughly comparable with that of the Federal Communications Commission in the United States. The CBC, however, continued its operating functions under its own president and Board of Directors.

Since 1952 the CBC had enjoyed a "monopoly" of television in the larger cities which it lost when in 1960 the new BBG began allocating channels to private enterprise in the same cities, thereby creating competition there between the CBC and private enterprise. At the same time the BBG tightened up its control over private radio and television stations by insisting that they pay greater attention to "public service" programming and to the employment of more Canadian talent on programmes. The BBG also laid down a requirement that 55 per cent of all programmes must be of Canadian content, and that the capital and control of these private stations must be mainly Canadian, to avoid any risk of their becoming satellites of American interests.

These changes created a situation for Canadian educators which was substantially different from that of the old days of radio when everything was done through the CBC, and the private stations played a largely passive part, as affiliates of the network. Now it was recognized that both the CBC and the private stations had an obligation to place suitable facilities at the educators' disposal for the development of educational television. Of course, this obligation did not solve the financial problem involved.

CANADA'S APPROACH TO EDUCATIONAL TELEVISION

Between 1954 and 1958 (when the new Broadcasting Act was passed), the CBC, at the suggestion of the education authorities, took the initiative in experimenting with school television. The first approach came from the Canadian Education Association which wrote to the CBC in 1953 inquiring what steps it proposed to take, and what machinery it would employ, for ensuring that the educational possibilities of the new medium would be adequately explored and developed. Behind this inquiry lay a certain nervousness, arising from the knowledge of the tremendous impact television was making on the minds and tastes of the rising generation of viewers. No one could deny that the new medium of "mass communication" held within itself some possibilities of serious damage to standards of value, as well as possiblities of great benefit through increased diffusion of knowledge and enlightenment. But the President of the Canadian Education Association in 1954, Dr. G. E. Frecker (Deputy Minister for Education, Newfoundland), had no doubt what the answer to these fears should be. In his presidential address to the Association in September he said:

Whether we like it or not, we are here in the middle of the twentieth century, with its wonders and dangers. Surely the sensible thing, instead of deploring the situation, is to take the more positive attitude of harnessing the marvellous forces which the age places at our disposal, harnessing them for the better education not only of our youth but of the people generally, and making them valuable allies in our efforts to give the growing generation the best our civilization affords, so that they may be not a disrupting element, but a bulwark.

The CBC replied to the CEA by referring the question of educational television to the National Advisory council. In considering this matter the Council was influenced by the rapid progress that was already being made in both Great Britain and the United States. The BBC had first decided to go ahead with establishing a regular television service to British schools, while in the United States the Federal Communications Commission had reserved channels (Very High Frequency and Ultra High Frequency) for the exclusive use of educators, who were already planning to use them for the setting up of a number of educational television stations.

Accordingly, early in 1954 the National Advisory Council recommended to the CBC a policy of experimentation in two directions: first, telecasts to be viewed at home by children, as a visual supplement to the national school broadcasts heard in the class-room, and secondly, a series of school telecasts specifically planned for school viewing. The first recommendation was carried out in March 1954 but on a very small scale. Four thirty-minute programmes were presented, in out-of-school periods, to supplement the national school radio series "Life in Canada Today" (grades 5–8). About 6,000 pupils who had heard the radio programmes also viewed the telecasts. A large majority reported to their teachers in favour of continuing these visual supplements to their radio lessons. A report entitled *Can TV Link Home and School?* was printed, but no further action was found to be feasible.

FIRST NATIONAL EXPERIMENTS

Meanwhile, in November 1954 the first experimental series of class-room telecasts was conducted jointly by the Council and the CBC. The aim was "to determine whether, and to what extent, television could help the teacher in her daily class-room work." The programmes were planned by the Council's Television Committee, under the chairmanship of Mr. George Croskery, Secretary-Treasurer of the Canadian Teachers' Federation. The Committee decided to limit the age range of students to two levels, grades 5–6 and grades 7–8. No programmes were aimed at either the primary or at the high school grades.

Eight half-hour programmes were given, in early afternoon periods, four for each grade level of student. The subjects were, for grades 5-6, social studies (history), health (safety), art, and literature (children's books); and, for grades 7–8, social studies (history), social studies (geography), science (conservation), and social studies (current events). The telecasts were produced in Toronto, and heard over a CBC network of 16 stations. They were received in 513 class-rooms (nearly 18,000 pupils) in eight provinces. The programmes were evaluated by 287 teachers of grades 5–6 and 226 teachers of grades 7–8. Of these, 32 per cent rated the value of school telecasts high as a teaching aid, 62 per cent medium, and 6 per cent low. Twenty-eight per cent requested a regular schedule of school telecasts for the future; 65 per cent asked for a further experiment in 1955–56; and 7 per cent urged that school telecasts be dropped. Teachers rated the social studies programmes (history and current events) as the best.

The viewing conditions in the class-room were generally regarded as satisfactory, but many problems were raised in connection with the curriculum, such as the difficulty of fitting programmes into an already crowded teaching schedule, the necessity of rearranging the daily time-table to fit in television lessons and of altering curricular order to fit in with the telecasts, the time used in preparing the class and doing "follow-up" work, and the time lost in moving classes from one room to another for viewing purposes.

In a number of the teachers' evaluations, reference was made to educational films in terms suggesting that school telecasts were liable to duplicate such films and were therefore a waste of time. Pupils' reactions to the programmes were generally enthusiastic.

The general conclusion reached in the report on this experiment (published by the CBC under the title *Television in the Classroom*) was that "Television programmes jointly planned and executed by teachers and broadcasters have a definite contribution to make as a teaching aid. However, further experiment will be required to clarify the precise nature and extent of this contribution." The report stressed the importance of publishing a detailed illustrated manual to assist teachers to make the most effective use of the programmes; it stressed also that "wherever possible television programmes should be viewed in a class-room rather than in an auditorium," and that television re-ceivers with at least a 21 inch screen be used. Finally, the CBC was requested to keep in mind the possibility that educational authorities might at some future date request the establishment of regular service on national school telecasts.

These recommendations were duly submitted to the CBC which decided to carry out a second and clarifying experiment to investigate the class-room uses of television for those student grades and in those curriculum subjects not covered by the previous experiment; to continue experimentation in the areas of social studies (especially current events) and science which seem to have the greatest potential usefulness; and to explore further the relationship between instructional films and educational telecasts as class-room teaching aids.

Suggestions for suitable programme topics were received from teachers in all parts of Canada. Their suggestions (numbering 428) were duly examined for technical feasibility and expense, and the following were adopted: three programmes (ten minutes each) on language for grades 2–4; three programmes (thirty minutes each) on science and social studies for grades 2–4; three programmes (ten minutes each) on social studies, geography, for grades 4–6; three programmes (twenty minutes each) on science for grades 4–6; and three programmes (thirty minutes each) on art, science, and social studies for grades 7–10. These 15 programmes were presented in April and May 1956 over a network of 29 stations (6 CBC-owned and 23 privately owned), in afternoon periods between 1:45 and 3:30. Some of the programmes were produced "live," others on kinescope in order that they could be heard in all ten provinces. A 36-page multigraphed manual giving full details of each programme was made available to all teachers using the telecasts. The programmes in this experiment were viewed in 721 Canadian schools, the number of classes viewing the telecast being 1,841, the number of students , 62,450. In most cases the schools used television receivers installed on a rental basis.

More than four-fifths of the teachers evaluating the programmes in this second experiment rated the value of television as a teaching aid as "medium" or "high." Approximately one-quarter asked for the establishment of a regular schedule of school telecasts while two-thirds asked for further experimentation. These figures corresponded closely with the findings of the first experiment and provided strong evidence of the value of television as a teaching aid and the desire of teachers to have more of it. Significantly, about 12 per cent of the teachers specifically urged that future programmes should be linked still more closely with the curricula of individual provinces.

Social studies and science headed the list of the programme subjects that were best liked by the teachers and were most recommended for future school telecasts.

On the question of films *versus* television, the comments of the teachers were indefinite. The great majority expressed no opinion on the subject but of the small minority who did express themselves more had a slight preference for films over telecasts. It appeared from the evidence that further consideration might usefully be given in the future to a closer correlation of the use of films and telecasts, and the television medium for distributing suitable films and film strips to the class-room.

The Canadian Home and School Federation conducted a national survey of parent opinion of this second Canadian experiment in classroom television which showed that parents reacted favourably. Thirty-seven per cent of them asked that school television be "put on a regular basis as soon as possible," while a further 51 per cent recommended a continuation of the experimentation. The report on this second experiment was published by the CBC under the title *School Television in Canada.*

LOCAL EXPERIMENTS

Halifax

Following the two national experiments, educational opinion began to turn in the direction of local experiment. In the fall of 1957 a special committee of the Directors of the Canadian Education Association recommended that further experimentation should take place at the provincial and interprovincial level. The CBC declared itself ready to co-operate through its regional centres, but stipulated that, to avoid overlapping and duplication of expense, all experiments involving the use of its facilities should be co-ordinated through the National Advisory Council.

The question of cost-sharing now came to the front. In provincial school radio broadcasting as we have seen, the CBC had been accustomed to bearing the indirect costs of programmes planned by provincial departments of education, while the departments themselves bore the direct costs. However, this simple formula, if applied to television, would produce quite different results. The indirect costs of television were proportionately much greater than the direct costs; moreover, for television purposes the CBC had adopted a system of cost accounting which seemed to many educators to inflate these costs on paper to a very high figure. Consequently, both parties felt alarmed at the possible liability they might incur by continuing the old formula. It was agreed that no clear-cut decision on the sharing of television costs could be reached without considerable further practical experience on both sides. As an interim formula, it was suggested that local experiments could be conducted on the following basis. The CBC would list in detail a basic minimum of facilities (covering studios, camera work, rehearsals, costuming and make-up, graphics and designs, etc.) which it would agree to place without charge at the service of the educators. Anything beyond this basic minimum would be charged to the educational authority conducting the experiment. This formula was provisionally adopted by both parties and remained in force for the next two years.

The first experiment under the new formula was tried in the city of Halifax in January 1958. The Halifax Board of School Commissioners decided to investigate the possibilities of using television lessons in the Halifax schools and to report on the results. The Board's plan was supported by the Nova Scotia Department of Education.

Nine half-hour programmes were presented by the Board jointly with the CBC Maritime region. Three were in grade 4 science (electricity), three in grade 6 geography (Nova Scotia), and three in grade 8 mathematics. The telecasts were presented over CBC station CBHT Halifax in an early afternoon period. They were viewed in 27 city schools, with 3,019 pupils participating. A report was prepared from questionnaires filled in by the teachers concerned.

The technique of presentation by "master-teacher" was used. Three teachers of recognized ability in the class-room were chosen, and afterwards assigned the television topics they were to teach. Unfortunately the teachers were neither experienced in the grade level at which they were called upon to teach nor specialists in the field to which they were assigned. Consequently they were not able, in the short time available for preparation, to make full use of their capacities on television. The best of the three series was that on electricity for grade 4, given by Mr. A. G. MacIntosh.

Dr. R. E. Marshall, the Superintendent of Schools, in a report made to the Halifax school board, came to the following conclusions:

A lesson which may be very effective when presented in the classroom may not secure the same satisfactory results on television. The successful television teacher must not only be experienced and competent in classroom teaching, but must also be qualified to translate the effective classroom technique into an equally effective television technique. . . . To hold completely the attention of the pupils, the television teacher needs something of the personal magnetism of the successful actor.

Highly qualified and efficient teachers seem to be more interested in securing television lessons designed to supplement and enrich the regular work of the classroom than they are in the presentation of "master-teacher" plan lessons which follow closely the course of studies as laid down by the Provincial Department of Education.

Manitoba

Manitoba was first among the provincial departments of education to experiment with school television and to repeat its experiments on a regular basis, largely due to the energy and enthusiasm of Miss Gertrude McCance, the Department's Supervisor of School Broadcasts, and to the keen co-operation of the officials of the CBC Prairie region.

The telecasts were broadcast over two stations in the province and were viewed in 117 schools, with a pupil audience of 18,301. Most schools used either borrowed or rented receivers.

The first experiment, in February and March 1958, consisted of two series of three half-hour programmes addressed respectively to senior high school and junior high school classes. The senior programmes were in chemistry, physics, and biology; the junior programmes were on arts and crafts, natural science, and English literature. They were planned by teacher committees and presented as demonstration lessons, reviews, interviews, and dramas. Exceptional success attended the English literature programme, "The Elizabethan Theatre" (grades 7, 8, and 9), which included a performance of the Pyramus and Thisbe scenes from *A Midsummer Night's Dream*. This programme later received an award from the Columbus (Ohio) Institute for Education by Radio and Television.

The Department's report on the experiment concluded that "educational television can serve a useful purpose in rural as well as urban centres" and that "some teachers are prepared to make use of it." School telecasts ought to employ both the "master-teacher" technique and "dramatization by professionals." It recommended that the scope of the programmes should gradually be extended to improve their quality and that they should be planned more closely to fit into the needs of the school curriculum. Stress was laid on the importance of using teachers as consultants or on committees.

This first experiment was followed a year later, in February and March 1959, with a second and more ambitious experiment. Six programmes were again given, in short series all aimed at the junior high school level. There were three social studies (history) programmes, two English language programmes, and one mathematics programme. All were planned "to stimulate the work of even the best qualified teachers." They were again broadcast over two television stations in the province in early afternoon periods, and were viewed in 69 schools with a student audience of 23,465.

Among the Department's conclusions, based on teachers' reports, were that "television is a practical teaching aid for schools in Manitoba" and that "within Manitoba are the resources necessary to produce good school television programmes." However, the Department stressed that the programmes must be of a professional quality, that "the teacher-narrator is essential and the studio class a distraction," and that "dramatization, where it can be used, is the most effective technique." Television need not be a passive means of instruction; it "cannot replace the teacher in the classroom, but can provide enriching and supplementary material."

In 1960 Manitoba did no further experiment of its own, but took part in a project of the four Western provinces. In 1961, however, Manitoba resumed action on a provincial basis, with a series of programmes presenting Sean O'Casey's drama *Juno and the Paycock* in half-hour instalments.

In its first two experiments, Manitoba made use of the facilities provided by the CBC in its Winnipeg studios, according to the cost-sharing formula accepted in 1957. However, the formula was found, after experience, to have some shortcomings. It proved difficult to hold the CBC's share of costs down to the basic facilities listed in the formula. Accordingly, a revision was worked out during 1959, to last until 1963–64, by which the CBC, broadly speaking, agreed to bear wholly the costs of national school telecasts, and to bear the indirect costs of municipal, provincial, and regional school telecasts, while leaving the educational authorities to bear the direct costs. The revised formula stressed the importance of early consultation between the CBC and the educators at all stages of scripting and production of programmes.

Closed-Circuit Experiments

The Scarborough (Ontario) Board of Education was the first to try the use of closed-circuit television, in November 1957 at Winston Churchill Collegiate Institute. "Closed circuit" involves the use of a cable system of communication over which school telecasts can be transmitted within a local area. It does not involve the use of the facilities of the CBC or privately owned television stations. The conclusion was that this type of school television could be justified for lessons that required the use of special or costly scientific apparatus, or were of a type not usually taught in an average secondary school and would be difficult to repeat. It was stressed that teachers participating in television teaching would need special training and that some students and teachers would have to be trained in the care and operation of equipment.

In March 1960 Scarborough conducted a second closed-circuit experiment at the David and Mary Thompson Collegiate. The resulting report recommended that all new secondary schools in Scarborough should have conduits built in, for later installation of coaxial cable for closed-circuit television, and that provision should be made for a studio of class-room size and for originating programmes in the school auditorium, science laboratories, and other class-rooms. The report also listed the special values closed-circuit television would have for secondary schools for lessons and demonstrations by skilled teachers or other specialists; for close-up studies during a television lesson; for telecasting films, film strips, slides, and other audio-visual materials, including open-circuit telecasts from outside; for telecasting to overflow groups of students in viewing rooms, thus making repetition of programmes unnecessary; and for stimulating an interest in the arts. The report also recommended that "control of television as a means of communication in education be kept in the hands of teachers."

In 1958 the Faculty of Education at the University of Alberta conducted a closed-circuit television study in Edmonton. Lessons were given in social studies, science, mathematics, music, and art at grades 5 and 6 levels. At the same time Mr. R. D. Armstrong of the Edmonton Public School Board conducted an experiment in the field of work-study skills involving the use of maps, charts, and tables in social studies. Little difference appeared between the effectiveness of the teaching done by television and the same teaching done in the class-room. The Faculty of Education also studied the use of closed-circuit television in teacher training, with favourable results.

In the early months of 1960 the School Broadcasts Branch of the Alberta Department of Education presented a series of ten television lessons in science (electricity and magnetism), for grades 7 and 8, using facilities provided by the local private station CFRN-TV. The specific aims of the series were to present the students with factual information and to encourage them to set up their own experiments where possible. The programmes were of a "master-teacher" type, given by an Edmonton school principal. The results were held by the Department of Education to be satisfactory and encouraging.

Newfoundland

In September 1959 the schools of Newfoundland were closed for three weeks on account of a polio epidemic. The privately owned television station in St. John's, CJON-TV, approached the Newfoundland Department of Education with an offer to use their facilities to fill the educational gap, insofar as the schools of the Avalon peninsula

(8.6 per cent of the whole school population of the province) were concerned. At the time the CBC had no television station in St. John's and no television network covering the province.

The established teaching aids of radio, films, film strips, and slides which were widely used in Newfoundland schools were powerless because the children were out of school. The Department of Education, therefore, being anxious to provide help to the children right in their own homes, readily accepted the offer of station CJON-TV. Within 48 hours a series of television lessons had been planned, to be given by eleven volunteer teachers. The series lasted for three weeks, from September 1 to 18, during which time 130 twenty-minute lessons were presented in ten subjects ranging from "Arithmetic for Beginners" to "English Literature for Grade 11." The lessons were presented in both morning and afternoon periods; at the end of the second week short tests were given on the work covered in seven subjects.

The technique of presentation included direct teaching, dramatization, interviews, films, and film strips. Naturally the time available for organizing the programmes was too short to avoid many faults. The lessons were in fact necessarily improvised and the teachers largely untrained in television techniques. Nevertheless, according to the Department, "the effects of this television series have been felt in every field of Newfoundland school life, even in adult life."

The Newfoundland Teachers' Association undertook to evaluate the series with the help of parents and teachers in those parts of the province reached by the telecasts. A questionnaire was distributed through the teachers in twenty-three St. John's schools to the homes of their pupils. Parents were asked to have their children view the telecasts and answer the test questions. Subsequently (December 1959) the Association published its findings in a research bulletin. The bulletin stressed the point that "these findings are not to be taken as too basic for similar television programmes as . . . they were hastily conceived and even more hurriedly carried out." Both teachers and parents, in their replies to the questionnaires, pointed out that "programmes would have to reach a higher and more consistent level of educational planning, production and performance before they could be considered as a satisfactory counterpart to the school broadcasting service." Many teachers, parents, and pupils failed to take advantage of the opportunity of viewing the telecasts, but those who did formed a favourable, if critical, opinion. Over four-fifths of the parents who replied to the questionnaire considered that the series had been successful in interesting children in school work. Nearly two-thirds of the teachers replying agreed with this opinion.

The Association concluded its survey by expressing the hope that the Department of Education would follow this initial experiment with a further effort, more carefully controlled and sustained, in the field of educational television. So far this hope has not materialized.

Ontario

Until 1961 the Ontario Department of Education showed little interest in school television. Initiative in the matter was, therefore, left to individual school boards in the large urban areas. In 1959 a number of educational and cultural organizations in the Greater Toronto area joined together to form the Metropolitan Educational Television Association of Greater Toronto which included the boards of education of thirteen municipalities in Toronto, the University of Toronto, the Art Gallery, the Public Libraries of Toronto, the Royal Ontario Museum, and other community agencies.

META, as the new organization was called, aimed at the encouragement of the use of television for educational purposes, particularly within the Toronto broadcasting area. It offered to function as a clearing house for information, and to co-ordinate the uses of the existing CBC and forthcoming private television stations in the area in matters of educational television. It offered also to give consultation service and to evaluate programmes. In November 1959 it presented a brief to the newly established Board of Broadcast Governors asking for the reservation of VHF and UHF channels for educational television purposes, looking forward in particular to the ultimate establishment in Toronto and other centres of educational television stations.

In 1961 META secured financial help for the first time from the Ontario government, together with the offer of experimental production facilities from the Ryerson Institute of Technology, in Toronto. META entered into active negotiation with CBLT and the new private station CFTO to present educational television programmes in the metropolitan area. It also launched a television training course for teachers and others at the Ryerson Institute in the summer of 1961.

About the same time as META became active, the Ontario School trustees' Council began to display increased interest in educational television, and set up in 1959 the Ontario Educational Television Association, which also presented a brief to the BBG, and undertook to evaluate the CBC's national series of school telecasts. The OETA aimed to serve as a rural couterpart to META, and in particular to look after the interests of the non-metropolitan parts of the province and to ensure that the benefits of educational television were made available to the rural as well as to the urban schools of the province.

Toronto. In 1960 the Toronto Board of Education began to take an interest in educational television. It began by presenting over CBC's local outlet CBLT a series of thirteen telecasts on Thursday mornings from January to the end of March. The telecasts were produced on colour film in the studios of the Board's Teaching Aids Centre under its director, W. Bruce Adams, but of course were presented by television in black and white. Three subjects were chosen: four on guidance for grade 9, four on art for grades 6–8, and five on science for grades 3–5. The programmes were viewed by 111 classes from 77 elementary and 16 secondary schools, and evaluated through 171 reports received from teachers, inspectors, and administrators. The majority reported that the broadcasts presented well-chosen materials in appropriate sequence and that they aided the class-room teaching of the subjects chosen. The absence of colour on the telecasts and the compression of visual material on the small television screens were both criticized. The camera work was praised.

In April 1960 the Toronto Board conducted another experiment, this time with closed-circuit television, at Riverdale Collegiate Institute. Three groups of pupils viewed the same series on guidance, one as produced on colour film, the other as presented by television, while the third group of pupils viewed both. No significant difference in quantity of learning, retention, or attitudes between the first two groups was reported, but the third group retained significantly more information than the others.

The Western Provinces

A Saskatchewan Educational Television Council was formed in 1958 to promote the development of the work in that area. However, lacking the necessary facilities and funds, this body has confined its work to collecting information and preparing the ground for future action.

In November 1960 the four Western provinces presented jointly with the CBC (Prairie and Pacific regions), two short series of telecasts at the intermediate and senior elementary level. The programmes were planned by the Western School Regional Committee, but prepared by the Saskatchewan and British Columbia departments of education, neither of which had previously presented any telecasts. Saskatchewan prepared four half-hour telecasts on the history of mathematics for grades 5–8, "to demonstrate that the development of mathematics was an outgrowth of local needs," which were produced in Winnipeg. British Columbia prepared four half-hour telecasts on physical geography for grades 7–9 which were produced in Vancouver.

All eight telecasts were presented in an early afternoon period over a network of stations covering the four provinces. No complete report on the results of the series has yet been published.

NATIONAL SERIES RESUMED

All these local experiments served to demonstrate the flexibility and variety of the possible approaches to school television. They also underlined the factor of high cost, which seemed likely to inhibit the education authorities from making any rapid increase of experimentation or regular provision of programming. Alberta and Manitoba emerged as the two provinces most likely to go forward on a regular but modest basis of presenting telecasts as teaching aids in their schools.

Under these circumstances the National Advisory Council favoured resuming school telecasting on a national basis, with the object of providing a sufficient service of programmes to encourage schools to equip themselves with television receivers on a permanent basis, instead of relying on temporary borrowings and rentals.

In 1960 and 1961 the CBC, on the Council's recommendation, undertook to present a thirteen-week series of half-hour national school telecasts, presented once a week in early afternoon periods from January to April. The 1960 series comprised six short series: for grades 2–3, four fifteen-minute programmes on music ("Rhythm and Melody") and four fifteen-minute programmes on social studies ("Children of Other Lands"); for grades 4–6, four fifteen-minute programmes on geography ("The Face of Canada") and four fifteen-minute programmes on general science; and for grades 7–9, five ten-minute programmes on current events and five twenty-minute programmes on history ("Where History was Made"). The 1961 series consisted also of six short series: for grades 2–3, four fifteen-minute programmes on social studies ("Homes of Long Ago"); for grades 4–6, four twenty-minute programmes on geography (transportation) and four fifteen-minute programmes on natural science (animal life); and for grades 7–10, five ten-minute programmes on current events and five ten-minute programmes of personality interviews. Five twenty-minute programmes on history ("Where History was Made") were also presented.

Evaluation of the 1960 series was left mainly to individual departments of education, teachers, and school trustees. The Ontario Teachers' Federation reported that (subject to some criticism by teachers on content and class-room discipline) interest was effectively aroused and recall of knowledge effectively assured by the telecasts. This judgment was confirmed by a similar but separate evaluation undertaken by the Ontario School Trustees' Council.

The national series (and a local science series over CFRN-TV in Edmonton) was also made the subject of a survey by the Alberta Federation of Home and School Associations to define the attitude of parents to school telecasts and to the future use of television for educational purposes. Eleven hundred questionnaires were sent out, of which over one-third were returned completed. The general reaction of parents was favourable to educational television in general, but many referred its use in adult education, or for supplementing information in the class-room, rather than for giving definite instruction leading to credits. The local science series was preferred to the national series.

In the school year 1961–62 the national school television service expanded considerably. There were 58 half-hour programmes covering two days a week through the winter, as follows: for senior high school, nuclear physics (four thirty-minute programmes), Canadian writers (four thirty-minute programmes), Shakespeare's *Macbeth**(five thirty-minute programmes); for ages 13 and upwards, Dickens' *David Copperfield** (serialized in five instalments), Canadian history* (three thirty-minute programmes on W. L. Mackenzie, Lord Durham, and Lord Elgin), music* (six thirty-minute programmes), and the United Nations (four thirty-minute programmes); for ages 10–15, "Ten Minutes With" (four ten-minute programmes interviewing well-known Canadians) and current events (eleven ten-minute programmes); and for ages 10–13, social studies (five thirty-minute programmes, "Here and There"), elementary biology (four twenty-minute programmes), geography (four twenty-minute programmes), exploring nature* (five thirty-minute programmes), the weather (five twenty-minute programmes), physical education (five fifteen-minute programmes), and the Eskimo (five fifteen-minute programmes). The asterisked programmes were on films adapted from American, British, and Canadian sources.

PRE-SCHOOL EDUCATIONAL TELECASTS

Two of the most succesful CBC contributions to educational television have been at the pre-school level. As we have seen, the success of "Kindergarten of the Air" led to a demand for a similar type of programme on television and the result was the establishment in 1957 of "Nursery School Time," a daily quarter-hour programme produced partly in Montreal, partly in Toronto. Its chief aim, as set forth by an advisory committee of nursery and kindergarten experts, is "to assist in the development of young children through programmes that are both educational and entertaining." Topics are chosen to acquaint the young child with new and interesting aspects of life about him, in his home and his community, and occassionally in the world beyond these limits. Songs, stories, music, and handiwork all play their part in developing these topics. The programmes are given by a nursery school teacher, aided by children and puppets.

The same techniques have been applied since 1958 to the teaching of French. According to Dr. Wilder Penfield and other experts, the pre-school age is the best time of life for the child to assimilate a second language, without undergoing the pains of formal study. "Chez Hélène" is a daily fifteen-minute telecast designed to develop the child's capacity to understand fluent French speech, even before he has learned to read and write. The Tan-Gau method of oral French, which is based on the repetition of speech patterns and language units, is used in this series. On the television screen, the teacher (Hélène Baillargeon) always expresses herself in French, but the children who accompany her respond at first in English. Then gradually, as Dr. Gauthier says, "the ear loosens the tongue" and the children viewing the telecasts begin to use French phrases and short sentences.

"Chez Hélène" was devised to meet an urgent need in Canada to increase bilingualism among both its French-speaking and its English-speaking population. The merit of the programme is its adaptability to the aquisition of either language through the oral method. English-Canadian families follow the telecasts to learn French; French-Canadian families follow it to improve their English. The programme is enriched (like "Nursery School Time") with live actors, puppets, and various graphic devices. Both "Chez Hélène" and "Nursery School Time" enjoy large audiences in all parts of Canada, although the time of broadcasting (early afternoon) is admittedly not ideal for the young people.

PROBLEMS OF CANADIAN SCHOOL TELEVISION

Although there has been this variety of experimentation in Canada during the past seven years, the rate of progress of educational television in this country has been much slower than in either Great Britain or the United States. In Great Britain, both the BBC and commercial television have undertaken responsibility for a regular service of school telecasts throughout the school year. Each now produces two hours of television programming five days every week. On the whole, the telecasts aim to enrich the school curriculum and to supplement the teacher's capacity, rather than to give direct teaching. In the United States, thanks largely to generous grants made by the Ford Foundation for the Advancement of Education and other educational foundations, nearly sixty educational television stations have been built and brought into operation, and a pool of programme material provided through the National Education Television and Radio Center. Universities, colleges, school boards, and community agencies have all contributed to this development, which provides not only enrichment courses, but a great deal of direct teaching through television. Many colleges and universities now provide credits for students taking recognized television courses. The closed-circuit experiments now being conducted at Hagerstown (Maryland) and numerous other centres are also primarily concerned with direct teaching by television. In addition, state-wide networks of educational television stations are being set up, and the ambitious Mid-West Airborne Television project, covering an area of 125,000 square miles, has been successfully launched. These are but a few of the immense variety of experiments that have been, or are now being, conducted in the United States. They are buttressed by an elaborate apparatus of research, whose evaluation of the pedagogic results achieved has been, in general, highly favourable.

The reasons for the slower rate of progress in Canada are not hard to find. Canada has not Britain's advantage of a high concentration of population within a small territorial area. Nor does she have a system of education which lends itself to centralized exploitation of the television medium. On the other hand, Canada does not possess the great financial resources which are available in the United States for promoting all kinds of experimentation in educational television. In particular, few Canadian foundations are prepared to give the kind of stimulus provided in America by the Ford Foundation. Again, Canada does not suffer from the acute shortage of teachers which has persuaded so many American educators to turn to direct teaching by television.

Canadian teachers have so far shown a preference for using telecasts for "enrichment" purposes. They fear lest its use for direct teaching may lead to a weakening of the professional status of the class-room teacher, and an undermining of the personal relationship between pupil and teacher. Possibly the rapid expansion of student enrolment in schools which is in prospect for the coming decade may alter their thinking on this matter. Recent American research studies have emphasized that direct teaching by television has special value in accelerating the learning process among backward pupils.

Another factor that has influenced teachers' opinions is the rivalry between film and television in education. Many education authorities have made a large capital investment in films, film strips, and film projectors; they therefore ask, "Can television do anything for the school that film cannot do equally well?" There is no direct answer to this question, except to point out that both are audio-visual aids, that each has its own function to perform, and that the changing techniques of production and distribution are making the issue increasingly unreal.

In Canada there has been growing co-operation between the two governmental agencies, the CBC and the National Film Board, in the field of television. For some years past, the National Advisory Council and the CEA-NFB Advisory Committee have co-operated in the field of educational television. The NFB has made many television programmes on film for the CBC for network broadcasting and arrangements have now been made to make these films also available for school television. One of the most successful of the programmes in the first CBC school television experiment ("House of History—A visit to Mackenzie House") was made on film by the NFB. It was also noted that nearly one-third of the programmes in the 1961 schedule of national school broadcasts were produced on film.

THE FINANCIAL PROBLEM

The main reason for the slow growth of Canadian school television is finance. The CBC has lately announced that it cannot continue indefinitely to bear the lion's share of the cost of school television and has urged the education authorities to spend more money and employ more staff on its development. The private commercial television stations are still less able to afford to subsidize telecasts. The educators themselves, except for the wealthier urban school boards, are not yet disposed to spend large sums on television for they are not convinced that expenditure on educational telecasts is as urgent as expenditure on class-rooms, equipment, and teachers' salaries.

On this matter Canada's experience compares more closely with that of other small countries, rather than with such major powers as America, Britain, and Russia. Canada must find some cheaper way to achieve comparable results.

AMERICANIZATION

An alternative way of developing school television in Canada is to draw much closer to the United States and to make greater use of the financial and programming resources of American television educators in Canada. On the one hand, there are many Canadian educators (especially in the universities) who feel that "education knows no national boundaries," and that there is therefore no reason why American programmes, produced on film or videoscope, should not be widely used for showing to Canadian schools, particularly in the fields of science (high school and college level) and literature. On the other hand, the CBC (as indicated earlier in this paper) came into existence for national reasons, and justifies its position by promoting Canadian educational and cultural programmes. In many provincial departments of education, too, there is a definite feeling in favour of relying on native resources rather than on material imported from either Britain or the United States.

The Board of Broadcast Governors, in addition to its attempt to foster "Canadianism" in television by prescribing a minimum percentage (55 per cent) of Canadian content for both the CBC and the privately owned television stations, has also required both parties in the industry to give time on the air to educational telecasts, and has urged the education authorities to increase their financial contribution to the use of the new medium in the class-room.

NATIONAL EDUCATIONAL TELEVISION CONFERENCE

At its 1960 meeting the National Advisory Council decided to arrange, with the help of the Canadian Education Association, the University of Toronto, and META, a representative national conference of educators and broadcasters to discuss the future development of educational television in Canada. The conference was held at the end of May 1961 in Toronto. Over one hundred persons attended, including six deputy ministers of education, six directors of curriculum, seven directors of education in large urban centres, a number of university teachers, school administrators, and trustees, representatives of the Canadian Teachers' Federation, members of META, and officials of the CBC, the National Film Board, the Canadian Association of Broadcasters, and the Board of Broadcast Governors.

Before the Conference, thanks to a special grant made by the Ford Foundation for the Advancement of Science, fifty-six of those attending were enabled to visit educational television centres in the United States and Britain; they were asked to report on their experiences at special group meetings in the Conference. Several leading experts in educational television from the United States and Britain attended the Conference by invitation and addressed its sessions. Notable among these were the addresses given by Mr. Kenneth Fawdry, Head of BBC School Broacasting (television), and Dr. Charles A. Siepmann, Professor of Education at New York University. They presented strongly contrasting views of the function of educational television. The British speaker stressed chiefly the "enrichment" value of school telecasts while the American stressed the urgency, in view of the critical shortage of teachers, of using television for "direct teaching." Addresses were also given by Dr. Andrew Stewart, Chairman of the Board of Broadcast Governors, and by Mr. Eugene Hallman, CBC Vice-President, Programming. Both stressed the urgency of greater financial provision, by the education authorities, for the development of educational television.

Seven Resolutions

At the end of three full days' discussion, the Conference passed seven resolutions, which asked the Board of Broadcast Governors to reserve VHF (Very High Frequency) and UHF (Ultra High Frequency) channels for the use of educational authorities; suggested that the provincial departments of education and the National Advisory Council should apply to the government under the terms of the federal Vocational Education Act, 1960, for a grant to alleviate unemployment by developing plans for educational television; asked the National Advisory Council to initiate future meetings of producing agencies, educational authorities, teaching organizations, and organizations representing performers and writers to help solve educational television problems; urged the CBC and its affiliates to make morning periods available for educational television programmes, as from September 1961; requested the National Advisory Council to establish a committee representing local and provincial authorities, broadcasting agencies, universities, and national educational organizations to deal with all phases of educational television; drew the attention of the National Advisory Council and the educational authorities to certain specifications for class-room receivers (including provision for the reception of VHF and UHF telecasts) drawn up by the Electronics Industries Association of Canada; and expressed appreciation of the CBC's services over the years to the development of Canadian nationhood and viewed with alarm any attempt to lessen the importance of the educational contribution made by the publicly owned radio and television networks.

SOME PROBLEMS

It is, of course, too early to asses the results of the National Conference, or the effectiveness of its resolutions. If carried out, they will probably stimulate the progress of educational television in Canada.

Large-scale, permanent progress in this field, however, depends on the solving of a number of fundamental questions that have been raised. The first is that of finance. The television service as a whole is an increasingly heavy financial burden on Canada, with its relatively small population, as is reflected in the federal government's policy of deflecting as much as possible of the responsibility for the television service onto the shoulders of private enterprise (advertising), and limiting the allocation of resources from the public purse to the CBC. The CBC then is forced to press the educators of Canada to assume a larger share of the costs of school telecasts.

To the school authorities the costs of professionally produced school telecasts seem very high, particularly as expressed by the CBC in terms of "cost accounting." The authorities wish to have proved that substantially increased expenditures on television are justified, in comparison with their expenditures on other educational needs. Moreover, the school radio broadcasts, which have been built up over the years by the CBC and the departments of education jointly involve only modest costs to both parties, yet provide a valued service to approximately three-quarters of the schools. Only a small minority of schools possess television receivers, and the number of programmes available to them has been small. If the school authorities agree to increase their expenditures on school television substantially, will they continue to use the facilities of CBC and private television stations (on an agreed cost-sharing basis), or will they seek to develop their own facilities (through establishing educational television stations on VHF or UHF channels, or through setting up closed-circuit facilities)?

Under existing conditions, the technical nature of television production means that the educators, while called upon to find a larger share of the costs, have less control over production than they are accustomed to have in school radio—at a time when it is more vital than it ever was in radio that the programmes should exactly meet curriculum needs and teaching requirements in the class-room. Though developing their own facilities would assure the educational authorities of absolute control over the production of the programmes, it would involve them in much greater capital outlay and running costs. The bigger cities could afford it, but not the rural areas to whom educational television means so much. The problem of control is not solved merely by solving it for the larger cities.

The financial question is closely allied to the functional problem involved. Will "enrichment" or "direct teaching" emerge as the prime function of educational television? The teachers have yet to make up their minds on this issue, and unless and until they do, the school authorities are unlikely to embark on the large expenditures referred to. There must also be some clarification of the respective roles of television, film, and other audio-visual aids in education. Can we envisage educational telecasts produced in the first place on film (rather than live in the television studio), then televised (i.e., distributed by television to the schools on a one-occasion basis), and subsequently made available to individual schools for projection and repeated study? Many of the existing objections to the ephemeral nature of school telecasts would thus be overcome. We may note that many television programmes are now being produced on video-tape, and that it is only a matter of time before video-tape recorders become cheap enough for purchase by individual schools.

The speed of technical improvement over the whole field is such that many educational authorities in Canada will wish to avoid committing themselves until the prospects are clearer. For example, the introduction of "teaching machines" (now being extensively developed in the United States) may have significant bearing on direct teaching by television.

Another matter of vital importance is teacher-training. How can class-room television develop on sound lines unless adequate provision is made in teachers' colleges, normal schools, and so on, for training new teachers in the uses of the television medium? No such provision has yet been made. If it were, it might involve far-reaching modifications of the present curriculum of teacher-training institutions.

The third problem facing Canada is the need for more effective machinery for co-ordinating the various interests concerned in the development of educational television. The National Conference of

1961 recommended enlarging or remodelling the National Advisory Council so that it could function as a kind of "parliament" for the whole movement, with status (and presumably financial resources) independent of the CBC. Canada would stand to benefit from co-ordinated planning of educational television development, and to enlarge or remodel the Council seems the only way—apart from the unlikely possibility that federal grants for educational television might be secured—to overcome existing financial handicaps, secure pooling of resources between the provinces, and ensure that a high standard of Canadian-produced programmes be maintained.

Bibliography
and Index

Bibliography

ALBERTA DEPARTMENT OF EDUCATION. *Alberta School Broadcasts.* Annual publication.

BRITISH COLUMBIA DEPARTMENT OF EDUCATION. *British Columbia School Broadcasts.* Annual publication.

CANADIAN BROADCASTING CORPORATION. *Five Years of Achievement—School Radio.* 1941.

——— *Radio-Collège Program Booklet* (in French). Annual publication, 1941–55.

——— *Young Canada Listens.* Annual publication, 1941–55.

———*Can Television Link Home and School? Report of a CBC Experiment.* 1954.

——— *Television in the Classroom—Report of a Canadian Experiment.* 1954.

——— *Canadian School Telecasts, January-April 1960: A Teacher's Guide.* 1959.

——— *Candian School Telecasts, January-March 1961: A Teacher's Guide.* 1960.

——— *Calling Young Canada—A Teacher's Guide to School Broadcasts in Canada.* 1961.

CANADIAN BROADCASTING CORPORATION AND MARITIME (ATLANTIC) DEPARTMENTS OF EDUCATION. *Atlantic School Broadcasts Manual.* Annual publication from 1945.

CANADIAN EDUCATION ASSOCIATION. *Proceedings of the National ETV Conference May 1961.* Published as a special issue of *Canadian Education and Research Digest* (1961).

CASSIRER, HENRY R. *Television Teaching Today.* UNESCO, 1961, 183–96.

EDWARDS, C. E. "School Radio Broadcasts Home Listening Survey," *Canadian Home and School Journal* (February 1956), 28–31.

LAMBERT, R. S. "Classroom TV in Canada," *National Association of Educational Broadcasters Journal* (October 1957), 10, 11, 23.
—— "The National Advisory Council on School Broadcasting," *Canadian Education* (1952).
—— "Music in School Broadcasting." In SIR ERNEST MACMILLAN (ed.), *Music in Canada*. Published by the University of Toronto Press in association with the Canadian Music Council, 1955.
—— *Radio in Canadian Schools*. School Aids and Textbook Publishing Co., Toronto, 1949.
LAMBERT, R. S. and BRONNER, R. "Inter-Commonwealth School Broadcasting—An Experiment," *BBC Quarterly* (Winter 1952–53), 210–15.
MANITOBA DEPARTMENT OF EDUCATION. *Young Manitoba Listens*. Annual publication from 1946.
—— "Manitoba School Television," *Manitoba School Journal* (February and September 1958).
McCANCE, GERTRUDE. "More Adventures in Television," *Manitoba School Journal* (September 1959).
NATIONAL ADVISORY COUNCIL ON SCHOOL BROADCASTING. *School Television in Canada—Report of a Further Experiment, April-May 1956*. Canadian Broadcasting Corporation, 1957.
NEWFOUNDLAND DEPARTMENT OF EDUCATION. *Newfoundland School Broadcasts*. St. John's, 1961.
ONTARIO DEPARTMENT OF EDUCATION. *Elementary School Radio Broadcasts*. Annual publication from 1945.
—— *Secondary School Radio Broadcasts*. Annual publication from 1946.
SASKATCHEWAN DEPARTMENT OF EDUCATION. *Young Saskatchewan Listens*. Annual publication from 1946.
UNIVERSITY OF TORONTO. *The Role of Television in Canadian Education: Proceedings of the National Educational Television Conference*. Toronto, 1961.

MIMEOGRAPHED OR MULTIGRAPHED MATERIAL

CANADIAN BROADCASTING CORPORATION. "Let's Take a Look." Teacher's Guide to the Use of the First Experimental CBC Television Programmes for Use in the Classroom, November, 1954.
—— "Now Let's Watch." Teacher's Guide to the Use of the Second Series of Experimental Canadian School Telecasts, April-May, 1956.
CANADIAN TEACHERS' FEDERATION. "Summary Report on Radio in Canadian Schools." 1956. Pp. 75.
—— "Survey of Radio in Canadian Schools." C.T.F. Research Study No. 1, 1956. Pp. 179.
CORBETT, E. A. "Report on Radio Broadcasting to Schools in Canada." Canadian Broadcasting Corporation, 1939.
ROSE, MARY JANE, "School Broadcasting in Canada." Ph.D. dissertation, Ontario College of Education, 1950.
RESEARCH AND EVALUATION SUB-COMMITTEE FOR THE ONTARIO EDUCATIONAL TELEVISION COMMITTEE. "An Evaluation of Canadian School Telecasts, 1960." Ontario School Trustees' Association, 1960. Pp. 13.

Index